Technological Advances in Condensed and Dried Milk

Dr. J. David

M.Sc. (Ag.); M.Sc. (Dairy Tech.); D.Phil. (Dairy Tech.);
P.G.D.B.A.; B.D.; D.F.P.; D.R.L.
Asstt. Professor
College of Food and Dairy Technology
Faculty of Engineering
Allahabad Agricultural Institute
(Deemed University)
ALLAHABAD (U.P.)
INDIA

KITAB MAHAL

First Edition : 2007

ISBN : 81-225-0456-6

SELLING AGENTS

1. **Kitab Mahal Agencies**
22-A, Sarojini Naidu Marg,
Allahabad—211 001
Phone : 2423466, 2424154

2. **Kitab Mahal Distributors**
28, Netaji Subhash Marg,
New Delhi—110 002
Phone : 23273230, 41563309
Fax : 011 : 23289285

3. **Kitab Mahal Agencies**
Ashok Rajpath
Patna – 800 004
Phone : 2670599, 2660531

Published by : Kitab Mahal, 22-A Sarojini Naidu Marg, Allahabad
Printed by : Century Printers, 22, Sarojini Naidu Marg, Allahabad

Preface

Grappling with the issue of shelf-life in milk and milk products when world pondered awhile, Condensed and Dried milk knocked at our doors as a major rescue, which established a new horizon in Dairy Industry today. A new and rather sophisticated technology in drying endowed with Membrane Processing which led us to Ultrafiltration (UF), Reverse Osmosis (RO) and Microfiltration (MF) created a niche amidst the plethora of scientific world that resulted in a new industrial revolution amidst dairy fraternity. Name them one by one, and one will find the fluorescence of technology in condensing and spray drying of milk and milk products running rampant around the globe.

A developing country like India, where rapid industrialization is taking place in every field since independence, it is the scientific knowledge that should percolate among the students, teachers and researchers in a specialized field like Dairy Technology. Contextual Indian books are very few that would cater to the needs of individual students. An area like condensing and drying of milk and milk products need to be focussed from Indian stand point of view. Both UG and PG students feel, the subject is too tough for them to grasp.

Inflorescence makes the bouquet arresting, so is the technology arranged in a lucid manner to touch the millions of readers, researchers and professionals. Therefore, the book, *Technological Advances in Condensed and Dried Milk*

Products is an obvious choice to satisfy their thirst for knowledge regarding the field of condensing and Drying.

It reveals many intricacies in the technological world which had so far been unreachable by common students. In a nutshell, it will serve as a handbook for every dairy technologist for his day-to-day use.

The author, out of his rich experience of two decades, presents the subject matter in a concise form in easy-to-read and easy-to-grasp language. The illustrations, diagrams and charts will help them as a path finder. The book will act as a Traveller's Tree, where every seeker of knowledge may quench his thirst by tapping it gently.

The author has taken every possible help from various journals, magazines, books and online articles for the preparation of the book. My sincere gratitude to all such authors, publishers, institutions for showing me the light. I have worked through their eyes. I owe you all. I have tried my best to acknowledge them in *Collateral Readings* cited at the end of the book. In case I missed any of them by an oversight, I seek their pardon. My teachers remain as light-house for preparation of the book, I salute them all.

I am a feeble person with human shortcomings. Without the abundant grace of the Almighty it was not possible to write this book in such a manner. All glory to God, sustainer of my life and activity ! I am also indebted to my parents for their constant prayers and good wishes for my every success in life.

The author specially owes a great deal to his wife, Dr. (Mrs.) Bipasha David, who, in the midst of family tempest, kept her cool in collecting valuable materials, arranging bibliography, helping in proofreading in a splendid way. My son Arpan Pushpit keeps on encouraging me by his innocent peculiar queries regarding the book. These are priceless asset for me !

Finally, I thank my well-wisher and family friend Mr. Indresh Kumar Agarwal of Kitab Mahal for his unstinted help to publish this book in a best possible manner.

J. David

Contents

manufacture and storage of milk powder—Milk fat, Milk proteins, Milk sugar, Milk salt; Shelf-life of milk powder; Uses of milk powder; Technology of whole milk powder production—Raw material, Standardization, Forewarming, Evaporation, Homogenization, Drying—Roller drying, Factors Effecting the quality of Roller-dried Milk Powder, Spray drying, Cooling, Technology of instant whole milk powder production; Instantizing affects the ability of dried powder particle; Instantization of whole milk powder—Process, Parameters in the production; Packaging Technology for skim milk powder production—Separation, Pasteurization, Heat treatment, Evaporation, Drying; Technology of instant skim milk powder production—Straight through method, Rewet method; Reconstituability and physical structure of different types of skim milk powder. Fortification of dry milk powder; Heat classified milk powder; Composition of milk protein and their specific properties; Denaturation of whey protein, Characteristics of denature whey protein, Effect of heat on protein, Whey Protein Nitrogen Index, Method for determination of WPNI; Categories of heat classified skim milk powder; Relationship between WPNI and solubility index; Heat treatment for manufacture of heat classified powder—Pasteurization, Pre-heating to evaporation, Effect of evaporator, Heating of concentrate, Effect of spray drying; Application of heat classified powder : Judging and Grading of whole milk powder & skim milk powder; Calculation; Reconstitution and Recombination of milk—Difference between Reconstituted and recombined milk; Basic reconstitution and recombining operation—Raw material handling, Milk powders, fats and oil, water, additives, Recombination of milk powder—Temperature and hydration time, Fat addition and emulsification, Air content, Heat treatment, Packaging and storage, Precautions/Notes, Discussion Questions.

List of Illustrations & Charts

List of Tables

Abbreviations

ADMI	American Dry Milk Institute
AMF	Anhydrous Milk Fat
AMUL	Anand Milk Union Limited
APEDA	Agriculture and Processed Food Products Export Development Authority
BA	Bifidus Activity
BIS	Bureau of Indian Standards
BOPP	Bioxially Oriented Polypropylene
BSF	Bifidus growth Stimulating Factor
CA	Cellulose Acetate
CAB	Cellulose Acetate-Butyrate
CITA	Centre for International Trade in Agriculture and Agro-based Industries
CPP	Cast Polypropylene
C_z	Sugar number/Sugar Index
DFHU	Direct Fired Heating Units
DMC	Direct Microscopic Count

EPS	Expanded Polystyrene.
ERH	Equilibrium Relative Humidity
et al.	And others
EVA	Ethylene Vinyl Acetate
EVOH	Ethylene Vinyl Alcohol
FAO	The Food and Agriculture Organization (of the United States)
FAPRI	Food and Agricultural Policy Research Institute.
FFS	Form-Fill-Seal.
GTR	Gass Transmission Rate
HDPE	High Density Polythene
HF	Hyper Filtration
HFFS	Horizontal Form Fill Seal
HIPS	High Impact Polystyrene
IS	Indian Standard
ISI	Indian Standard Institute
KJ	Kilo Joule
LDPE	Low Density Polyethylene
LLPD	Linear Low Density Polyethylene
MBRT	Methylene Blue Reduction Test
MVR	Mechanical Vapour Recompression
NDDB	National Dairy Development Board
NDRI	National Dairy Research Institute
NFD	Non Fat Dry
OPS	Oriented Polystyrene

PA	Polyamide
PE	Polyethylene
PER	Protein Efficiency Ratio
PET	Polyethylene trephtalate
PVC	Poly Vinyl Chloride
PVDC	Poly Vinylidene Chloride
RH	Relative Humidity
rmp	Revolution Per minute
RO	Reverse Osmosis
SCM	Sweetened Condensed Milk
SMP	Skim Milk Powder
SNF	Solid Not Fat
TFC	Thin Film Composite
TS	Total Solids
TVR	Thermal Vapour Recompression
UHT	Ultra High Temperature
V/W	Volume/Weight
W/V	Weight/Volume
WMP	Whole Milk Powder
WPC	Whey Protein Concentrates
WPNI	Whey protein Nitrogen Index
WVTR	Water Vapour Transmission Rate

1

Condensed and Dried Milk Industry – An Overview

Pondering back to the last century when dairy farmer and consumer grappled with the low shelf-life of milk and milk products, a major rescue came with the advent of condensed and dried milk. Due to the preserving effect and low water activity, condensed and dry milk products are desirable in regions with unfavourable climatic conditions. The evaporation technique as the classic concentration method has proved its worth in the dairy market. All milk components after evaporation and drying retain almost all original nutritive value of milk, thus, making it quite popular. Now-a-days, some special powdered products based on milk components such as whey protein concentrate, casein derivatives and infant formulas creating much uproar in the dairy industry.

India is the world's biggest market for malt-milk based food products. Market has been growing in the range of 5 to 8% for the last ten years (Dhillon, 2005)

The dried milk industry manufactures nearly 1,85,000 tonnes of infant formula in India, representing approximately 3.8% of, total production (Thompkinson *et al.*, 2005)

HISTORY AND DEVELOPMENT

Evaporated and Condensed Milk

1809 – Nicholas Appert, a French Food Scientist, announced his discovery of a means of preserving milk (without the admixture of sugar or other preservatives). Method of preparation—The milk was condensed by heating it in an open vessel placed in a water bath over fire to about 2-3rd of its original volume. The condensed milk was strained, cooled and poured into glass bottles, filling them to the brim and corking them tightly. The bottles so filled and sealed were then held in a boiling water bath for two hours. The milk so treated was found to have very good keeping quality.

1835 - Vacuum Pan was first devised in this year in England.

1856 - The commercially successful manufacture of condensed milk was initiated by the American, Gail Borden, who is known as 'the father of the process of milk condensing'. In this year, he received a patent both from USA and England.

1856 - Borden manufactured sweetened condensed milk, sold under the famous 'Eagle brand'.

1884 - John Meyenberg and Louis Latzen, established the first factory to make unsweetened, evaporated milk in Highland, Illinois.

1884 - John Meyenberg was granted a basic patent on his invention of a Sterilizer.

1887 - John Meyenberg, was granted patent on his process of preserving milk. His process patents provided sterilization by steam under pressure at a temperature not exceeding 116°C (240°F), while the sealed cans were continuously agitated by the revolving reel. This innovation provided the basis for a new Industry.

1961 - In India, the Amul Dairy is credited with the first-ever commercial production of sweetened condensed milk.

Dried Milk

13th Century - Marco Polo, reported that the soldiers of Kablai khan, a Mongol emperor, carried dried milk when on an excursion. He wrote that Mongols boiled milk, removed the cream and dried the remaining liquid in the sun. Before it was used, water was added to a portion of the dried material. Mixing was accomplished by the horse's movements while travelling.

1810 – Dried milk in tablet form by the French Scientist Nicholas Appert, by the air-drying of milk solids concentrated to 'doughty' consistency.

1855 – Grimwade of Britain developed a modified dry product from highly concentrated milk.

1855 – A patent was granted in England for a method of drying milk by evaporation in open pan.

1872 – Spray drying equipment invented by Percy of USA and Stauf of Germany (1901).

1889-1909 – Vacuum drum drier's were designed.

1902 – Roller dryer was invented by John A. Just of USA.

1903 – 1st Just–Hat marker roller dryer was installed in England by Cow & Gate Company.

1909 – First invention of a drum drier with two rolls.

1944 – Owing to the unfavourable reaction that many persons have to the term, skim milk, the dairy industry refers to dry skim milk as 'non-fat dry milk'. This term was incorporated into the Federal Pure Food, Drug and Cosmetic Act by Congress in this year. The term is descriptive of the product and does not minimize its high nutritive value.

1955 – In India, the first-ever commercial production of spray dried milk was started by AMUL Dairy, Anand. The first spray dryer was a single stage spray dryer imported from Niro A/s, Denmark. Today India has the most modern and automated milk powder plants with the large capacity of 60 TPD producing instant and agglomerated milk powders. The plants are locally manufactured with only critical components imported from abroad.

Malt-based Milk Food, Baby Food/Infant Milk Powder

1850 – An organisation known as Wiscousin Mellin Company manufactured first malt-based food product. This product was branded as 'Leigbig' extract and fed mainly to infants after mixing with sterile milk and water.

1883 – The process of manufacture of malted milk powder was invented by William Horlick of USA, at the request of physicians who wanted a babyfood prepared from milk and cereals.

1887 – The malted milk product was in the market. The convenience, nutritive value, palatability and digestibility of this product made it an instant success.

1930s – One of the pioneers was the British Company Cow & Gate Ltd., whose product became known in many parts of the world. In the records of the company is described the excitement caused by an urgent order in the 1930s from famous Maharajah in India for consignment of baby food powder. The company's export manager visualised the tremendous potential in India, but was somewhat disappointed when it later was learned that the powder was required for the Maharajah's racing stables.

1957 – In India, investigations on dried infant food were jointly undertaken by Central Food Technological Research Institute, Mysore and Amul Dairy, Anand.

1960 – A standardized process was evolved for the commercial production of Amul Infant/baby food from buffalo milk by the Roller Process.

1967 – Since 1967, Amulspray baby food has been in the market.

DEVELOPMENT OF INDIAN MILK POWDER PRODUCTION

India was primarily an import dependent country till early 70s. Most of the demand-supply gas of liquid milk requirements for urban consumers were met by importing Anhydrous milk fat/butter and dry skim milk or whole milk powder. The cheap imported skim milk powder, which was available then at a price of about Rs. 2,200 per tonne, was affecting adversely the production of milk and manufacturing of milk powder. The milk powder plants which were mostly in the private sector, remained underutilized. These plants produced mostly baby food for which were mostly in the private sector, remained underutilized. These plants produced mostly baby food for which were imported skim milk powder which was made available by the government. However, the canalization of imported skim milk powder quickly identified fake "Actual Users" and eliminated them. These fake "Actual Users" usually sold the imported powder at a very high price. As this source of clandestine sale of imported powder was eliminated, there was more demand for indigenous skim milk in the market which provided an incentive for the milk product factories with drying plant to manufacture skim milk powder. Milk powder production started rising. There was a temporary glut of milk powder in the market and the Government of India was persuaded by the milk powder manufacturers to remove the ceiling on the price of skim milk powder for the central pool of milk powder. The Indian Dairy corporation, which was then the canalizing agent, was permitted by the government to buy indigenous skim milk powder by open tender. This gave a boost to manufacture of milk powder in the country and milk powder production, and, therefore, milk conservation rose significantly since the work of canalization of milk powder started (Fig. 1.1). The pooling of the skim milk powder helped the indigenous producers to sell milk powder at a price which meets their cost of production and, at the same time, the dairies get skim milk powder to meet their shortfall at a price lower than the market price of the indigenous milk powder (Raychandran, 1996).

With the outset of the Operation Flood Programme, the scenario dramatically changed and commercial import of dairy products came to a halt except occasional imports of very negligible quantities.

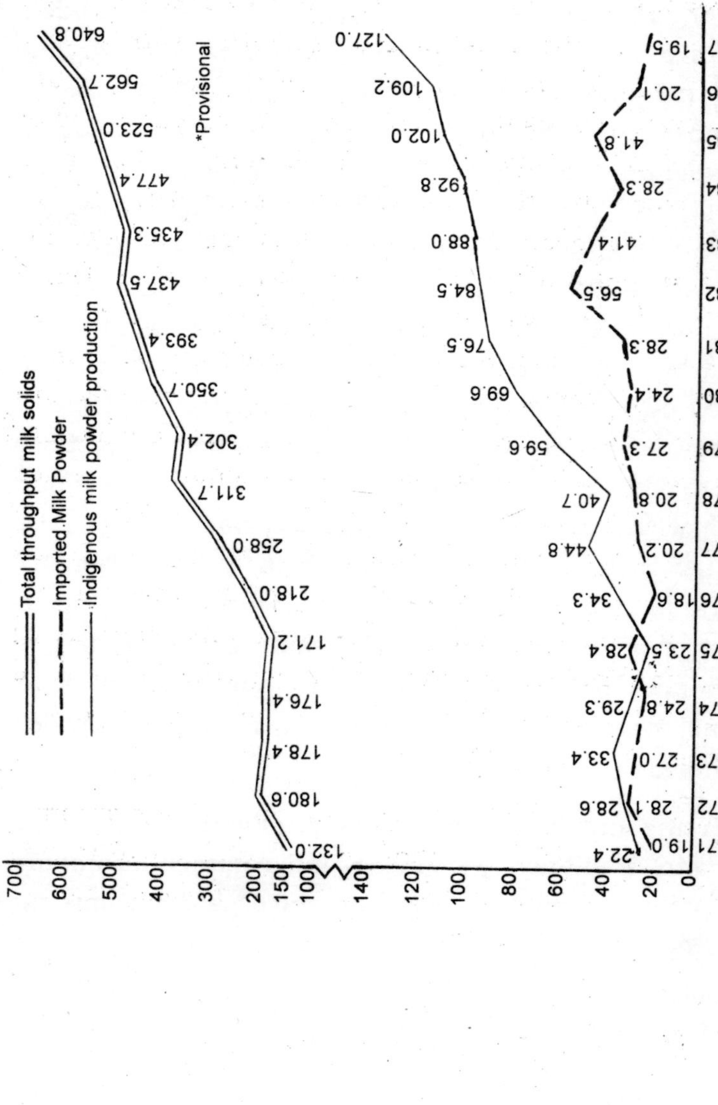

Fig. 1.1 Indigenous Milk Powder Production and Share of Imported Milk Powder in the Throughput of the Indian Dairy Industry

Source : Operation Flood – A progress Report by Indian Dairy Corporation.

In the 90s, India started exporting surplus dairy commodities, viz. skim milk powder, whole milk powder and ghee. These exports are in bulk and only occasional. The export of dairy commodity is regulated by the Agricultural and Processed Food Products Export Development Authority (APEDA) which fixes ceilings every year for exports. Such ceilings are fixed in consultation with the administrative ministry, after taking into consideration internal requirements and anticipated milk production. However, in 1999, the Government had issued a notification, removing the practice of fixing ceiling of export of milk products. The export scenario between 1993 and 1996 is shown in Table 1.1 (Banerjee, 1999). Table 1.2 shows imports and exports of SMP.

Table 1.1

Exports of Milk Powder against Ceilling (in Mt)

Year	Milk Powder
1993-94	2,609
(Ceiling)	(25,000)
1994-95	12,893
(Ceiling)	(25,000)
1995-96	2,975
(Ceiling)	(10,000)
1996-97	2,134
(Ceiling)	(10,000)

Source : APEDA

Table 1.2

Imports and Exports of SMP (Mt) (India)

Year	Imports	Exports
1997-98	670	1353
1998-99	1424	655
1999-2000	17252	2198
2000-01	547	3627
2001-02	48	14429
2002-03	17	11987
2003-04	9112	3070

MAJOR DOMESTIC PLAYERS

Major domestic players of dried milk (SMP, WMP, Dairy whitener) Malted milk food, Infant milk food is given in Table 1.3.

Table 1.3

Major Domestic Players of Dried Milk and Milk Products

Brand Name	Company	Products
AMUL	GCMF	Dairy Powder
Mother Dairy	Mother Dairy	Dairy Powder
Nestle	Nestle	Dairy Whitener, Infant Foods
Horlicks, Viva	Glaxo Smith Kline	Malted milk food
Boost, Maltova	Consumer Healthcare Ltd.	
Bournvita	Cadbury	Malted Food
Verka	Milkfed	Dairy Whitener
Britania Milkman	BNZF	Dairy Whitener

GLOBAL TRENDS IN CONDENSED MILK PRODUCTION

World's production of condensed milk and world's top condensed milk production countries are given in Table 1.4 and Table 1.5, respectively.

Table 1.4

World's Production of Condensed Milk (2003-04)

Name of Continent	Condensed milk (thousand tonnes)
Europe	1845 (46.42)
North America	1104 (22.77)
Asia	558 (14.04)
Oceania	49 (1.23)
South America	355 (8.93)
Africa	64 (1.61)
World	3975

Table 1.5

World's Top Condensed Milk Production Countries (2003-04)

Countries	Condensed Milk production (Thousand tonnes)
USA	793
Germany	545
Netherlands	300
Russian Fed.	204
Thailand	179
U.K.	163
Mexico	158
China	114

GLOBAL TRENDS IN MILK POWDER TRADE

Milk powders include skim milk powder (SMP)/Non-fat Dry milk (NFD), whole milk powder (WMP), butter milk powder (BMP) and milk protein concentrates. Milk powder can be recombined into liquid milk, products, particularly in tropical climates where fresh milk supplies are not available. It is also used in bakery, confectionery and milk chocolates processed meats, ready-to-cook meals, baby foods, ice-cream, yogurt, health foods, reduced-fat milks and animal fodder.

Given higher returns from cheese production and a declining supply of fluid milk, some milk is diverted from NFD into cheese production.

Production of Non-fat Dry Milk/Skim Milk Powder

The United States is the largest producer of non-fat dry milk with a production of 0.63 Mio. Mts. during 2004. New Zealand's production (0.3 Mio. Mts) was about half the quantity compared to that of the US during 2004; however, the NFD production trend shows a continuous growth over the last few years. At the same time, Australia used to produce larger quantities of NFD as compared to New Zealand but has reduced the production of this commodity since 2002.

Table 1.6 shows World's production of SMP and Table 1.7 World's top ten SMP production countries.

Table 1.6

World's Production of SMP (2003-04)

Name of Continent	Skim Milk Powder (Thousand tonnes)
Europe	1817 (51.86)
North America	818 (26.7)
Asia	237 (6.76)
Oceania	562 (16.04)
South America	59 (1.68)
Africa	11 (0.31)
World	3504

Table 1.7

World's Top Ten SMP Production Countries (2003-04)

Countries	Skim Milk Powder Production (Thousand tonnes)
USA	699
France	339
New Zealand	306
Germany	302
Australia	256
Russian Fed.	252
Japan	170
Poland	150
Canada	98
U.K.	71

India is the largest producer of NFD among Asian countries and is also one of the leading producers in the world Market. As per the forecast of FAPRI, the quantity of NFD produced by India during 2005 (0.27 Mio. Mts.) is expected to be higher than that produced by New Zealand owing to an anticipated decline in the production of the latter.

Consumption of Non-fat Dry Milk

The largest producer of NFD milk, the United States is also a largest consumer of this product. The consumption of NFD by USA during 2004 was 0.62 Mio. Mts., which has almost doubled compared to that of the year 2000 (0.34 Mio. Mts.). Mexico is the second largest consumer of NFD with consumption of 0.3 Mio. Mts. during 2004; however, the consumption trend is more or less stable and has shown a negligible rise over last 3-4 years period.

India, on the other hand, has increased its consumption of NFD and is expected to surpass the consumption demand of Japan which shows a declining trend for this commodity. Part of the reason seems to be that Japan is shifting towards more valuable products like cheese and flavoured milk beverages whereas India's growing demand for a number of milk items seems to be satisfied by the NFD, which can be converted into a number of varied products by reconstitution. (Sardana & Dokras, 2005).

Exports of Non-fat Dry Milk

Non-fat dry milk Imports exhibit major changes on the export front as far as world market is concerned. New Zealand is one of the major exporters of NFD to the world market with 0.31 Mio. Mts. of export during 2004. New Zealand exports of NFD had shown a continuous increase during 2000-04. Australia has drastically reduced its exports of NFD to the world market. Australia used to export larger quantities of this commodity as compared to New Zealand during 2000 (0.25 Mio. Mts.), however, the exports during 2004 suggest a decline of almost 0.1 Mio Mts. The United States, on the other hand, has been able to increase her exports by almost 0.1 Mio. Mts. over the last 2 years period. NFD export share of selected countries during 2000-2005 is given in Figure 1.2.

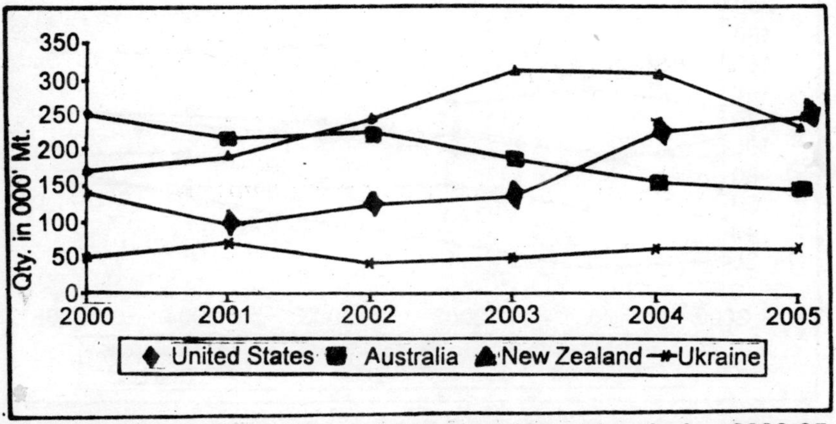

Fig. 1.2 : NFD Export Share of Selected Countries during 2000-05

Source : FAPRI Database

Imports of Non-fat Dry Milk

Major importers of NFD are the countries that have limited milk supply or production capabilities because of poor income, transport infrastructure, roads and lack of refrigeration facilities. Owing to high prices and limited supply, milk is perceived as a luxury product in such countries. Powdered milk is mainly imported to reconstitute it into various types of milk products to satisfy the routine demand for milk such as feeding infants, tea or other beverages.

Mexico is the largest importer of NFD in the world market with imports rising from 0.29 Mio. Mts. during 2000 to 0.7 Mio. Mts. during 2004. Indonesia and Philippines are other two major importers of this product which exhibit a constant rise in their imports of NFD

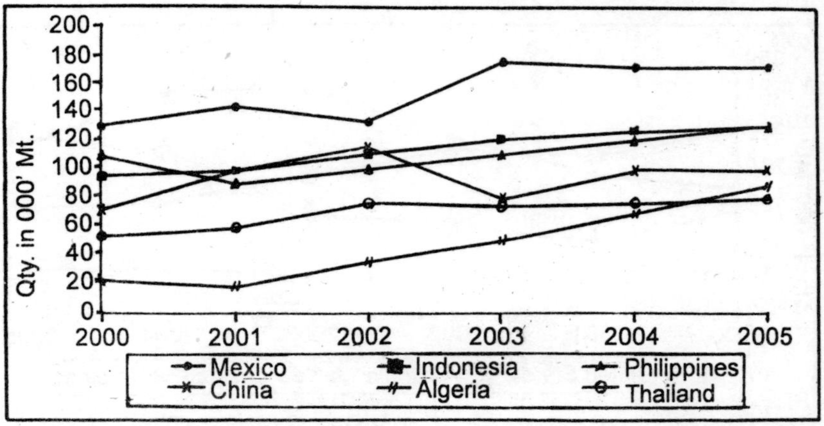

Fig. 1.3 : NFD Import Share of Selected Countries during 2000-05

Source : FAPRI Database

over the last four years. Interestingly, China, where milk is perceived as an infant food, had been very poor importers of NFD in the past. However, in recent years, China's NFD imports have showed a dramatic rise from 22,000 Mt during 2000 to 88,000 Mt in 2005. The growth trend for this commodity in China is expected to continue over next few years, which is going to increase the import demand for this commodity in the International markets. NFD import share of selected countries during 2000-2005 is given in Figure. 1.3.

Forecast of Milk Powder

Over the next decade, the world NFD production is expected to rise to 16.7% with greatest gains occurring in Australia, New Zealand, Russia and China. With the exception of the European Union, whole milk powder (WMP) production would grow in most countries, increasing a total of 24.3%. NFD and WMP prices rise annually by 2%.

As a by-product of butter, India's NFD production would grow 56.9%, creating excess supplies that allow NFD exports to rise to 17.2% annually.

South east Asia (Indonesia, Malaysia and the Philippines) is expected to increase its share of total NFD imports from 23% in 2004 to 34% in 2014. South east Asian WMP imports rise to 4.4% annually throughout the projection period.

DISCUSSION QUESTIONS

1. Chronologically describe the history and development of evaporated and condensed milk industry.
2. How was milk powder developed in India ? Discuss.
3. In a tabular form, explain the major domestic players in dried milk industry in India.
4. What is the global trend in condensed milk production?
5. Discuss the international milk powder trade.

2

Evaporated and Sweetened Condensed Milk

Gazing back to the eighteenth century, basic phenomena have been observed all over the globe concerning the shelf-life of dairy products. The dawn of nineteenth century experienced a sigh of relief with the appearance of evaporated and condensed milk in the dairy scenario. After the initial hiccups of introducing a new product by Gail Borden in America in 1850, the Borden condensed milk company grew faster and occupied the pivotal position in the dairy industry. Since then, condensed and evaporated milk industry has grown by leaps and bounds and became a household name in India, specially due to the sincere efforts of AMUL, who is taking the credit way back from 1961. At present, evaporated and condensed milk made their mark in the reconstitution of milk, for tea and coffee preparation, ice-cream manufacturing and several other milk drinks are at our doorsteps to energise us.

Condensed milks are the products attained by partial evaporation of water from whole milk or fully or partly skimmed milk, with or without the addition of sugar. It first came into being as a means to preserve milk solids.

Objectives

Condensed milks are prepared to meet the following objectives :

(i) To improve shelf life of milk

(ii) To save storage and package space

(iii) To serve as a resource material under scarcity conditions

(iv) To use as an additive for manufacture of ice-cream, candies and other food items

(v) To reduce cost of transportation.

TYPES OF CONDENSED MILK

A. Unsweetened condensed milk (evaporated milk) :

(i) Full cream sweetened condensed milk.

(ii) Unsweetened condensed skim milk.

B. Sweetened condensed milk :

(i) Full cream sweetened condensed milk.

(ii) Sweetened condensed Skim milk.

EVAPORATED MILK/UNSWEETENED CONDENSED MILK

Principles of the production of evaporated milk involve evaporation (removing a part of water from milk) and sterilization of the evaporated milk obtained, to provide better shelf-life of the product.

Desirable Attributes

Evaporated milk is a sterilized product, light in colour and with an appearance of cream.

Definitions

According to the PFA Rules, Unsweetened condensed milk (evaporated milk) is the product obtained from cow or buffalo milk or a combination thereof, or from standardized milk, by the partial removal of water. It may contain added calcium chloride, citric acid and sodium citrate, sodium salts of orthophosphoric acid and polyphosphoric acid (as linear phosphate) not exceeding 0.3 per cent by weight of the finished product. Such additions need not be declared on the label. Unsweetened condensed milk should contain not less than 8.0 per cent fat, and not less than 26.0 per cent milk solids.

Unsweetened condensed Skim milk (evaporated skim milk) is the product obtained from cow or buffalo skim milk or a combination thereof by the partial removal of water. It may contain added calcium chloride, citric acid and sodium citrate, sodium salts of orthophosphoric acid and polyphosphoric acid not exceeding 0.3 per cent by weight of the finished product. Such additions need not be declared on the level. Unsweetened condensed Skim milk should contain not less than 20.0 per cent total milk solids. The fat content should not exceed 0.5 per cent by weight.

Chemical Composition

The average chemical composition of evaporated milk is given in Table 2.1.

Table 2.1

The Average Chemical Composition of Evaporated Milk

Constituent	Content in 100 g Unsweetened Condensed Milk	
	Average	Range
1	2	3
Main components		
Water	74.0 g	73.7-74.4 g
Protein (N × 6.38)	7.0 g	6.49-8.40 g
Fat	7.74 g	7.57-7.90 g
Carbohydrates	9.76 g	—
Ash	1.50 g	
Salt Constituents		
Sodium	94 mg	67-152 mg
Potassium	297 mg	272-341 mg
Magnesium	—	—
Calcium	228 mg	213-243 mg
Manganese	—	—
Iron	0.13 mg	0.04-0.28 mg
Cobalt	—	—
Copper	0.035 mg	0.010-0.050 mg
Zinc	0.73 mg	0.50-0.90 mg
Phosphorus	213 mg	195-230 mg
Fluoride	—	—
Chloride	—	—
Iodine	—	—
Vitamins		
Vitamin A	0.090 mg	0.054-0.12 mg
Carotene	—	—
Vitamin D	0.1 mg	—

1	2	3
Vitamin E	—	—
Vitamin K	—	—
Vitamin B_1	0.067 mg	0.050-0.090 mg
Vitamin B_2	0.37 mg	0.32-0.44 mg
Nicotinic acid	0.20 mg	0.20-0.21 mg
Pantothenic acid	0.74 mg	—
Vitamin B_6	0.042 mg	0.030-0.053 mg
Biotin	2.8 mg	—
Folic acid	—	—
Vitamin B_{12}	0.10 mg	—
Vitamin C	1.4 mg	1.0-1.7 mg
Amino acids		
Isoleucine	0.45 g	0.30-0.58 g
Leucine	0.69 g	0.56-0.85 g
Valine	0.48 g	0.31-0.59 g
Methionine	0.17 g	0.11-0.27 g
Cystine	0.063 g	0.027-0.11 g
Phenylalanine	0.34 g	0.26-0.46 g
Tyrosine	0.36 g	0.28-0.44 g
Threonine	0.32 g	0.25-0.42 g
Tryptophan	0.098 g	0.067-0.13 g
Lysine	0.54 g	0.36-0.67 g
Histidine	0.18 g	0.085-0.34 g
Arginine	0.26 g	0.19-0.34 g

Source : Kessler (1988)

Chemical Standards

The chemical standards for evaported milk in India, USA and the UK is given in Table 2.2.

Table 2.2

Chemical Standard for Evaporated Milk in India, UK and USA.

Types of evaporated milk	Milk fat (%)			Total milk solids (%)		
	India	UK	USA	India	UK	USA
Full cream evaporated milk	<8.0	9.0	7.9	26.0	31.0	25.9
Skim unsweetened condensed milk	>0.5	—	—	20.0	20.0	20.0

Food and Nutritive Value

Evaporated milk rich in fat and fat soluble vitamins A, D, E and K, body-building proteins, bone forming minerals and energy giving sucrose.

Commercial Application

Evaporated milk is commonly used in nutrition, but usually not for drinking.

- It has a large market, for example, in tropical counties, at sea and for the armed forces. It is used where fresh milk is not available as a fresh milk substitute.
- In many countries where normal milk is available, it is used as a coffee whitener.

- Unsweetened condensed milk is also used as a substitute for breast milk. In this case, vitamin D is added.
- It is used as a component in various food preparations (Infant food, confectioneries, etc.).

TECHNOLOGY OF UNSWEETENED CONDENSED MILK PRODUCTION

The manufacture of unsweetened condensed (evaporated) milk consists of receiving and selection of raw milk, clarification, cooling, storage, first standardization preheating, evaporation, homogenization, cooling second or final standardization, packaging, sterilization and storage (Fig. 2.1).

Similar technology is utilized also when production of concentrated milk is based on recombination. In this case, standardization is taking place during recombination or reconstitution.

Receiving and Selection of Raw milk

It is necessary to pay special attention to the choice of raw materials. The quality of milk required for condensing is basically the same as that required for the manufacture of other milk products. The milk should be clean, sweet, free from offflavours and odour and reasonably free from extraneous material. The initial quality of raw milk has a direct bearing on the microbiological quality of the finished product. In order

Fig. 2.2 Flow chart of unsweetened condensed milk production

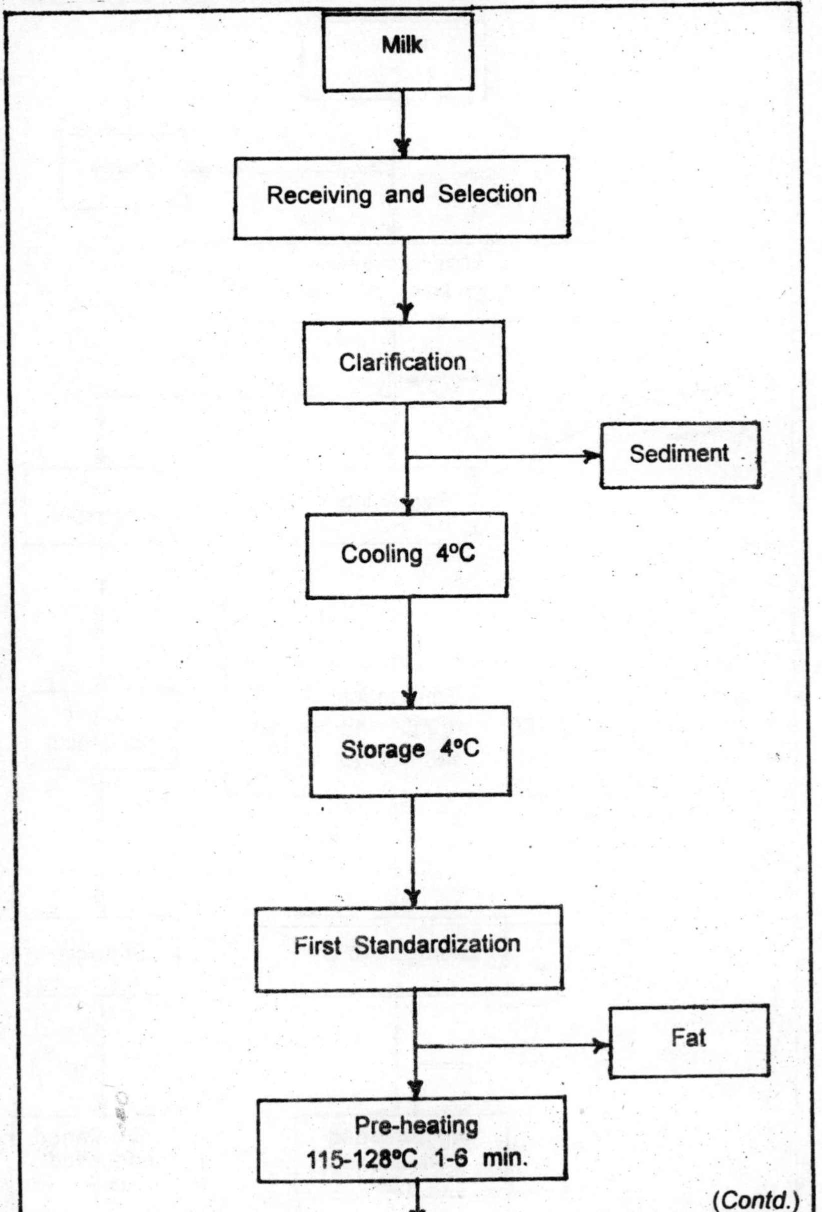

(Contd.)

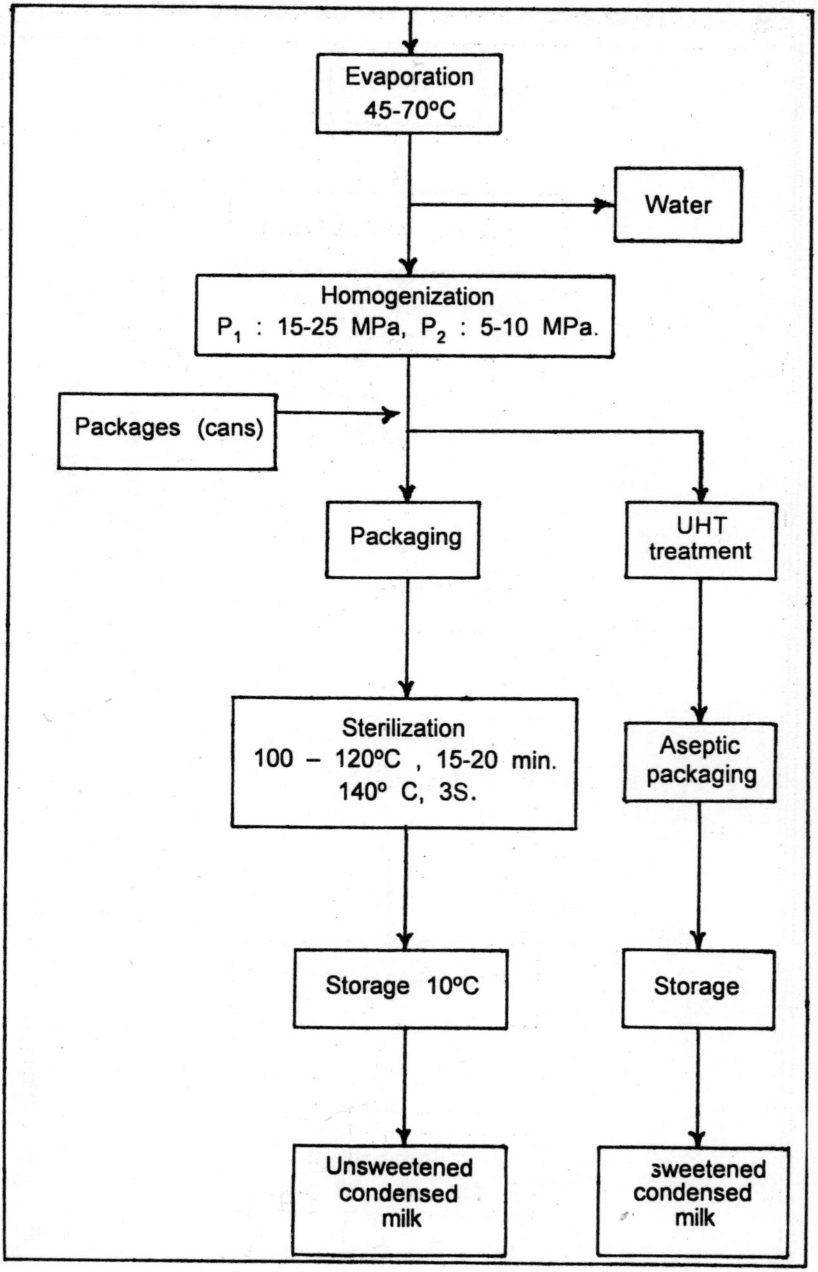
Evaporation
45-70°C
Water
Homogenization
P1 : 15-25 MPa, P2 : 5-10 MPa.
Packages (cans)
Packaging
UHT
treatment
Sterilization
100 – 120°C , 15-20 min.
140° C, 3S.
Aseptic
packaging
Storage 10°C
Storage
Unsweetened
condensed
milk
sweetened
condensed
milk

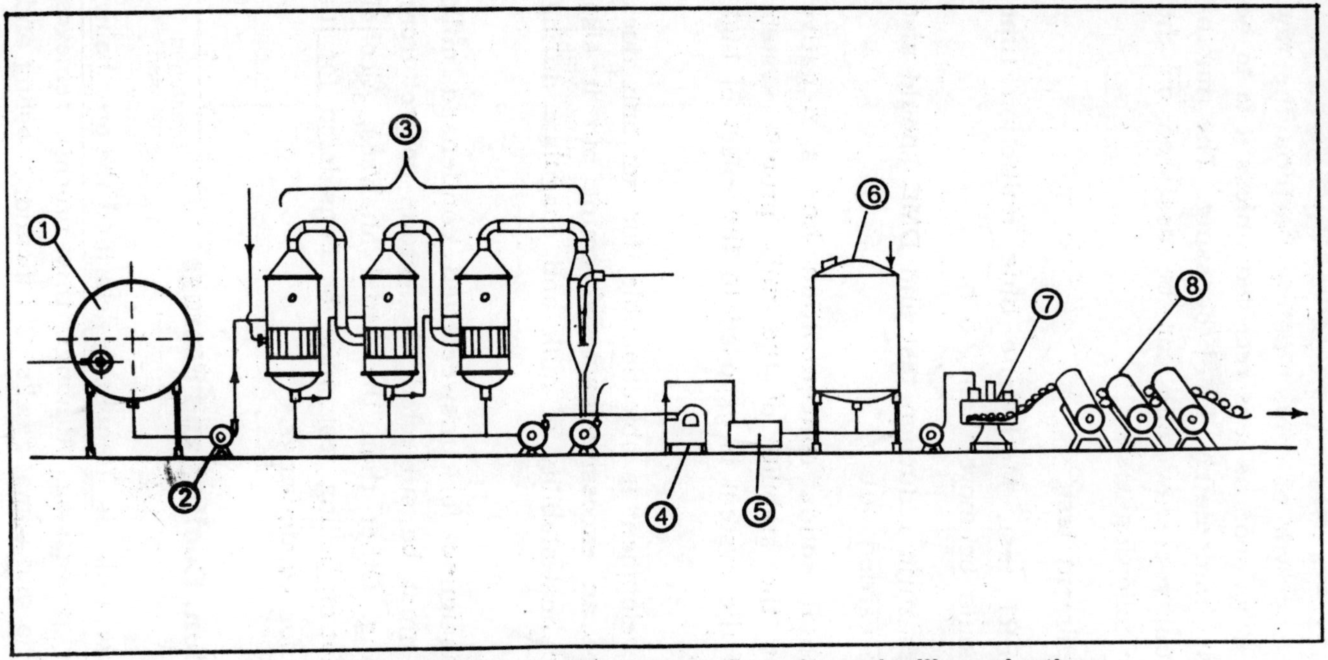

Fig. 2.1 Process line of unsweentened condensed milk production

1. Standardization tank, 2. Pump, 3. Three-effect vacuum evaporator, 4. Homogenizer, 5. Cooler, 6. Standardization tank, 7. Filling machine, 8. Continuous sterilizer.

to retard the growth of contaminating bacteria the raw milk is cooled as soon as it is received unless it is to be used almost immediately for processing. The micro-biological quality of raw milk can be assessed on the basis of the following tests :

(a) Sediment test.

(b) MBRT test, Methylene blue reduction time should be more than 2.5 h.

(c) Resazurin reduction test and DMC should also be carried out.

High total solids concentration has a negative influence on the stability of the milk protein system especially if the system is exposed to the effect of high temperature.

Acid development is objectionable, for not only does this indicate an excessive bacterial count but it also reduces the heat-stability of milk and coagulate during sterilization.

Milk assigned for unsweetened condensed milk production must be subjected to rigorous control from the beginning. Other than the usual analyses, Alcohol test and clot-on-boiling test must be considered for its acceptance for condensing.

Clarification, Cooling and Storage

The raw milk is frequently clarified before being diverted for processing. Although this process removes some bacteria and somatic cells. Clarification, cooling and storage of raw milk is done in the usual way; clarification by centrifugal separators, cooling by plate heat exchangers to 4°C and storage in tanks at the same temperature.

First Standardization

Standards for milk and airy products regulate the ratio of milk fat to non-fat solids in unsweetened evaporated condensed milk. The figures vary with the applicable standard, but are normally 7.5% fat and 17.5% solids-non-fat. This product is often called "double concentrated milk" (5%) total solids. Another common standard is "triple concentrated milk" with 33% total solids (often 4-10% fat content).

In the United States, Federal Standards for evaporate fat are not less than 25% of total milk solids in the final product (FDA, 1978). In the United Kingdom, fat to the non-fat solids ratio is 1 : 2 and in Germany 7.5 : 17.5 (Caric, 1990).

To achieve the desired ratio in the final product, the components are adjusted before evaporation. The ratio is achieved in this manner; the fat and SNF contents in milk are analyzed. On the basis of these values and desired ratio in the product, the value to which the fat content to be corrected is calculated by material balance.

The standardization of the ratio is most often done by separator-standardizers. However, if milk with high fat content enters the process at this stage of manufacture, a calculated amount of Skim milk powder is added. The same result is gained by mixing the initial raw milk with the calculated amount of Skim milk or by mixing milk with different fat contents. Standardization takes place in tanks designed for this purpose.

Modern automatic standardization systems permit continuous and extremely accurate standardization of both fat content and the relation between fat content and SNF of the basic milk.

Formulae

(i) $\% \text{ SNF in cream} = \dfrac{(100-\% \text{ fat in cream}}{(100-\% \text{ fat in milk})} \times \% \text{ SNF in milk}$

(ii) $\% \text{ SNF in Skim milk} = \dfrac{(\% \text{ SNF in whole milk}}{(100-\% \text{ fat in whole milk})} \times 100$

Pre-heating

Pre-heating decreases the total number of micro-organisms, thus intensifying the sterilization effect and enabling milder sterilization. But, the basic reasons for preheating are to increase milk stability during sterilization, and for its effects on the viscosity of the final product and this is extremely important for the product quality.

The effect of preheating on thermal stability

This is explained by the equilibrium that exists between acid and alkaline equivalents. The sum of the acid equivalents, P_2O_5, Cl, SO_3, CO_2, citrates, casein, albumin and globulin, is approximately equal to the sum of the alkaline ones, CaO, MgO, K_2O and Na_2O (0.1395 to 0.1397). The acid equivalents show a distinctly stabilizing effect on the protein system in milk, while alkaline equivalents, especially Ca^{2+} and Mg^{2+} performs opposite functions, leading to aggregation of the casein micelles and their destabilization and precipitation.

Equilibrium collapse, whether a consequence of disordered lactation or caused by calcium and magnesium, seriously endangers the stability of the protein system (Swais Good, 1986; Holt, 1985).

During heat treatment, a great part of the whey protein denatured, while calcium salts are precipitated. In this way, the protein complex of the milk is stabilized so that it can withstand subsequent sterilization, without coagulation taking place during the process or subsequent storage.

Preheating time and temperature

The heat treatment often integrated in the evaporation plant, takes place in a tubular or plate heat exchanger at a temperature of 93-100°C for 10-25 minutes or 115-128°C for 1-6 minutes (Caric, 1994), followed by chilling to about 70°C before milk enters the evaporator.

Evaporation

To avoid the undesirable changes of milk components caused by higher temperatures, evaporation is always conducted in a partial vacuum, thus reducing the evaporating temperature. This is based on the fact that the boiling point of a liquid can be lowered to correspond to a pressure below atmospheric pressure. To eliminate the growth of the microorganism, evaporation temperatures are never below 45°C. Staphylococci is developed at temperatures below 46°C. The heating medium is usually low pressure steam and indirect heat transfer is achieved through use of tubes, plates or falling-film type. Details of evaporator is given in chapter 3.

The evaporator is usually of multistage falling-film type. The milk passes through steam-heated tubes under vacuum. Boiling takes place at 65-70°C. The dry matter content of the milk increases as the water is boiled off. The density is checked continuously.

Homogenization

The concentrated milk is pumped from the evaporator to a homogenizer. In sweetened condensed milk processing, homogenization pressures may vary in the range of 15 to 25 mPa (150-250 bar) in the first stage and 5-10 mPa (50-100 bar) in the second homogenization stage.

The homogenization temperature in the production of unsweetened condensed milk is somewhat lesser than in ordinary dairy processing, matching the temperature of the evaporator, which most commonly is 50°C. Homogenization disperses the fat and prevents the fat globules from coalescing during subsequent sterilization.

Homogenization should not be too intensive that might decrease the stability of the protein, with the consequent risk of the milk coagulating during sterilization. It is, therefore, necessary to find the exact homogenization pressure that is high enough to produce the required fat dispersion, yet low enough to eliminate the risk of coagulation.

In UHT processing, homogenization during pre-treatment is normally low, in order to avoid separation during storage of the concentrate prior to the final heat treatment. The main homogenization then takes place in the UHT treatment.

Final Standardization and Intermediate Storage

Standardization

In the stage of processing, the ratio of milk fat to solids non-fat is adjusted. The total dry matter is standardized in already evaporated milk if the first standardization has been carried out. Standardization is achieved by the addition of water, skim milk, evaporated skim milk or homogenized cream.

Addition of stabilizing salt

The heat stability of the condensed milk can be improved by the addition of stabilizing salts, usually disodium or trisodium phosphate. The quantity of phosphate to be added is determined by sample sterilization. Tests are needed because variation occurs between batches of milk.

Intermediate storage

Test sterilization is time-consuming test-procedure and further processing must wait until the results are available. In the meantime, the concentrate must be stored. However, long-term storage should be avoided, not only to prevent bacterial growth, but long cold storage may increase the tendency of agglomeration.

Vitamin addition

Any addition of vitamins is also done at this stage.

Sterilization and packaging

After final standardization, the product is subjected to sterilization either before packaging or after packaging. Earlier, the practice was to fill evaporated milk in cans followed by sealing it hermetically and then sterilizing by heat. However, the process now has been replaced by first sterilizing the product followed by hermetically sealed packaging.

Canning and sterilization

Unsweetened condensed milk is usually packaged in cans of 0.4 kg to several kilograms which are then sterilized. Filling is carried out at low temperatures to avoid foam formation. Canning machines for condensed milk automatically fill and seal the cans before sterilization. Some 48-60 cans can be filled simultaneously, depending on the capacity of the machine.

The sterilization of filled, sealed cans is carried out continuously in can sterilizers at 100-120°C for 15-20 minutes (Fig. 2.2). The sterilizers are composed of preheating parts and two basic cylinders for sterilization and cooling. The interior contains rotating can holder. During this operation, the can is rotating not only around the cylinder interior, but also around its own axis. This provides uniform heat transfer and minimizes the adverse heat induced changes in the product. The first cylinder, where the sterilization is completed, is steam heated and cooling in the next cylinder is by cold water. That drops the temperature to 25-30°C. The can move continuously through the whole system by an elevator that regulates the sterilization to 15-20 minutes duration and cools it for about 15 minutes.

UHT and Aseptic packaging

Sterilization by direct or indirect heating prior to packaging at an ultra high temperature is known as flow sterilization.

In this case, following test sterilization and addition stabilizer, if required, the milk is pumped to the UHT plant, where it is heated to 122-140°C for a period of 4 second to 8 minutes.

The time/temperature combination of the UHT treatment as well as the homogenizing conditions (mainly conducted aseptically, but, sometimes, in both aseptic and non-aseptic mode) are to main extent determining the final product colour, viscosity and storage stability. After cooling, the unsweetened milk is packed in aseptic board packages and stored.

Risk of protein destabilization

The risk of protein destabilization is greatest by UHT indirect sterilization, less by in-can sterilization and least with UHT direct sterilization (direct contact between milk and the steam).

Storage

The quality of unsweetened condensed milk is altered during storage. The rate of change depends on the applied time-temperature regime.

Temperature

The evaporated milk goes brownish if the storage temperature is too high and protein may precipitate if the storage temperature is too low. It has been established that evaporated milk with higher dry matter content can be successfully stored upto a year without any significant quality change when stored at temperature of 6-8°C. Therefore, the desirable temperature for storage of this product should not exceed 10°C.

Stabilization salt

The better protein and fat stability and longer storage capability of unsweetened condensed milk is obtained with the addition of 4-5% whey powder and stabilization salts (Martynova and Belyaeva, 1979).

Relative humidity

Corrosion can occur in cans, so, the relative humidity in the air where they are stored should be below 50%.

Turning of cans during storage

Cans are turned frequently during storage, especially if kept for long period of time. Turning the can presents the product from forming a surface cream layer that if stored for long periods at low temperature.

Shelf-life of evaporated milk

UHT treated evaporated milk has normally a shelf-life of 6-9 months.

FROZEN EVAPORATED MILK

Freezing and thawing of evaporated milk brings about some changes in sensory qualities such as fat destabilization, setting of solids in the container and casein precipitation under drastic conditions. In some cases, freezing may distort the cans or weaken the can seam. Sometimes, sedimentation is observed in the frozen product due to the crystallization of lactose which redissolves with difficulty (Webb *et al.*, 1978).

UNSWEETENED FLAVOURED CONDENSED MILK

Unsweetened flavoured condensed milk is produced with Coffee, Cocoa or other flavouring. In production, the milk is first standardizated to 25.5% total solids and 7.8% milk fat, and heat-treated at 120°C for 1 minute. Stabilizing salts are added and the product is homogenized at 70°C. After the flavouring is added, the product is sterilized. The final product of coffee flavoured unsweetened condensed milk, for example, contains 2-4% dry matter of Coffee extract, which gives the product intensive Coffee taste and aroma. It has uniform consistencies and is stable for 6 months of storage (Antonov *et al.*, 1982).

Judging and grading

High grade evaporated milk should have mild pleasant flavour, a relatively viscous body and uniform smooth texture and uniform colour. The American dairy Science Association standard is furnished below as a guide for judging and grading of evaporated milk (Table 2.3).

Table 2.3

A Score Card for Judging and Grading of Evaporated Milk

Item	Perfect grade (%)
Flavour and odour	40
Body and Texture	35
Colour	5
Fat Content	10
Total Solids	10

Adulterants/preservative must be absent.

SWEETENED CONDENSED MILK

Sweetened condensed milk is a canned product and it contains sugar. It can be stored for longer periods in almost its natural state.

Desirable attributes

A good well made sweetened condensed milk (SCM) has uniform colour (cow milk : yellow colour of varying intensity; buffalo milk : slightly cream with greenish white tint); the surface and under- layer of both are free from mold growth, lumps, fat separation or scum. It has uniform consistency which is pourable like a ribbon that tends to level by itself with no sign of folds on the surface. Its texture is smooth, having absence of sediment and sandiness. Its flavour is clean, sweet representing a blend of added sugar and milk solids.

Definition

According to the PFA Rules : **Sweetened condensed milk** is the product obtained from cow or buffalo milk or a combination thereof, or from standardized milk, by the partial removal of water and after addition of cane sugar. It may contain added refined lactose permitted flavours, calcium chloride, citric acid and sodium citrate, sodium salts or orthophosphoric acid and polyphosphoric acid not exceeding 0.3% by weight of the finished product. Such additions need not be declared on the label. Sweetened condensed milk should contain not less than 9.0 per cent milk fat, not less than 31.0 per cent total milk solids and not less than 40.0 per cent cane sugar.

Sweetened condensed skim milk is the product obtained from cow or buffalo skimmed milk or a combination thereof by the partial removal of water and after addition of cane sugar. It may contain added refined lactose calcium chloride, citric acid and sodium citrate, sodium salts of orthophosphoric acid and polyphosphoric acid not exceeding 0.3 per cent by weight of the finished product. Such additions need not be declared on the label. Sweetened condensed skimmed milk should contain not less than 26.0 per cent of total milk solids and not less than 40.0 per cent cane sugar. The fat content should not exceed 0.5 per cent by weight.

Chemical Composition

The average chemical composition of sweetened condensed milk is given in Table 2.4.

Table 2.4

Average Chemical Composition of Sweetened Condensed milk

Constituent	Content in 100 g Sweetened condensed milk	
	Average	Range
1	2	3
Main Components		
Water	26.1 g	25.0-27.0 g
Protein	8.2 g	81-8.3 g
Fat	8.8 g	8.4-9.0 g
Carbohydrates	55.1 g	—
Ash	1.8 g	1.7-1.9 g
Salt constituents		
Sodium	88 mg.	84-92 mg.
Potassium	360 mg.	340-376 mg.
Magnesium	—	—
Calcium	238 mg.	223-252 mg.
Maganese	—	—
Iron —	ca 0.2 – 0.3 mg.	
Copper	—	—
Zinc	—	—
Phosphorous	236 mg.	228-250 mg.
Fluoride	—	—
Chlorine	—	—
Iodine	—	—
Vitamins		
Vitamin A	0.14 mg	0.11-0.18 mg
Carotene	—	—
Vitamin D	0.13 g	—
Vitamin E	—	—
Vitamin K	—	—
Vitamin B_1	0.094 mg.	0.090-0.097 mg.

1	2	3
Vitamin B_2	0.39 mg	0.35-0.44 mg.
Nicotinic acid	0.24 mg.	—
Pantothenic acid	0.82 mg.	—
Vitamin B_6	0.059 mg.	—
Bitonin	3.2 mm	—
Folic acid	—	—
Vitamin B_{12}	0.50 mg	—
Vitamin C	3.8 mg.	
Amino acids		
Isoleucine	0.52 g	0.35-0.68 g
Leucine	0.80 g	0.66-1.00 g
Valine	0.56 g	0.36-0.69 g
Methionine	0.20 g	0.13-0.31 g
Cystine	0.073 g	0.031-0.13 g
Phenylalanine	0.40 g	0.30-0.54 g
Tyrosine	0.42 g	0.33-0.51 g
Threonine	0.38 g	0.29-0.50 g
Tryptophan	0.12 g	0.079-0.15 g
Lysine	0.64 g	0.42-0.78 g
Histidine	0.22 g	0.10-0.40 g
Arginine	0.30 g	0.22-0.40 g

Source : Ktermeier and Lechner. (1973)

Chemical Standards

BIS and PFA standards is given in Table 2.5, and Table 2.6 shows American Standard and British Standards of Sweetended condensed milk.

Table 2.5

PFA and BIS Specification for Sweetened Condensed Milk

Constituent	PFA*		BIS**		
	SCM	SCSM	SCM	PSCSM	SCSM
Milk fat, % by wt.	9.0 min.	0.50 max	9.0 min	>3 to <9	0.50 max
Milk solids, % by wt. min	31.0	26.0	31.0	28.0	26.0
Sucrose, % by wt. min	40.0	40.0	40.0	40.0	40.0
Stabilizing salts, % by wt. max.	0.30	0.30	—	—	—
Titratable acidity (as LA) % by wt. max.	—	—	0.35	0.35	0.35
Bacterial count / g max.	—	—	500	500	500
Coliforms	—	—	—ve	—ve	—ve
Yeast & mould/g max	—	—	10	10	10
Coagulase+ve *staphylococcus aureus*	—	—	Absent	Absent	Absent
colour	—	—	whitish to light brown		
Accelerated storage test curdling38°C, 14 days	—	—	Absence of bulging, churning, No objectionable taste, odour and sliminess		

Source : *Anon (1993) ** IS 1166-1986.

Table 2.6

American Standard and British Standard of Sweetened Condensed Milk

Constituent	American Standard	British Standard
Fat content (%)	8	9
Milk solids–Non-fat, (%)	20	22
Lactose, (%)	10.3	11.4
Sucrose (%)	10.3	11.4
Sucrose (%)	45	43.5
Water %	27	25.5
Lactose, g/100 g water	38.3	44.6
Sucrose, g/100 g water	167	171
Concentrator factors Q	4.60	5.00

Source : P. Waltra *et al.* (1999).

Food and Nutritive Value

SCM has high nutritive value. It is rich in fat and fat soluble vitamins A, D, E and K, body building proteins, bone forming minerals and energy-giving lactose and sucrose.

Commercial Applications

SCM, because of its high total solids concentration, especially sugar, is not directly used in nutrition.

- Its primary household use is in the preparation of various food including coffee, ice cream, pudding and beverages.

- Food processing industries use SCM as a sweetened and as the source of milk components.
- In the confectionery industry, SCM is used as the best source of milk components in caramel candy processing.

TECHNOLOGY OF SWEETENED CONDENSED MILK PRODUCTION

Principles

The principles for the production of SCM involves evaporation followed by the addition of sugar to the concentrated milk to obtain a product with extended shelf-life. Sugar addition ensures prolonged shelf-life because of osmoanabiosis. With this process, a high concentration of sucrose or other sugar increases osmotic pressure which causes plasmolysis of microorganisms present.

Technology

The technological process in SCM milk production (Fig. 2.3) includes the following operations the receiving and selection of milk, clarification, cooling and storage, first standardization, heat treatment, evaporation, sugar addition, second standardization, cooling with crystallization and packaging and storage.

Fig. 2.3 : Flow chart of Sweetened condensed milk

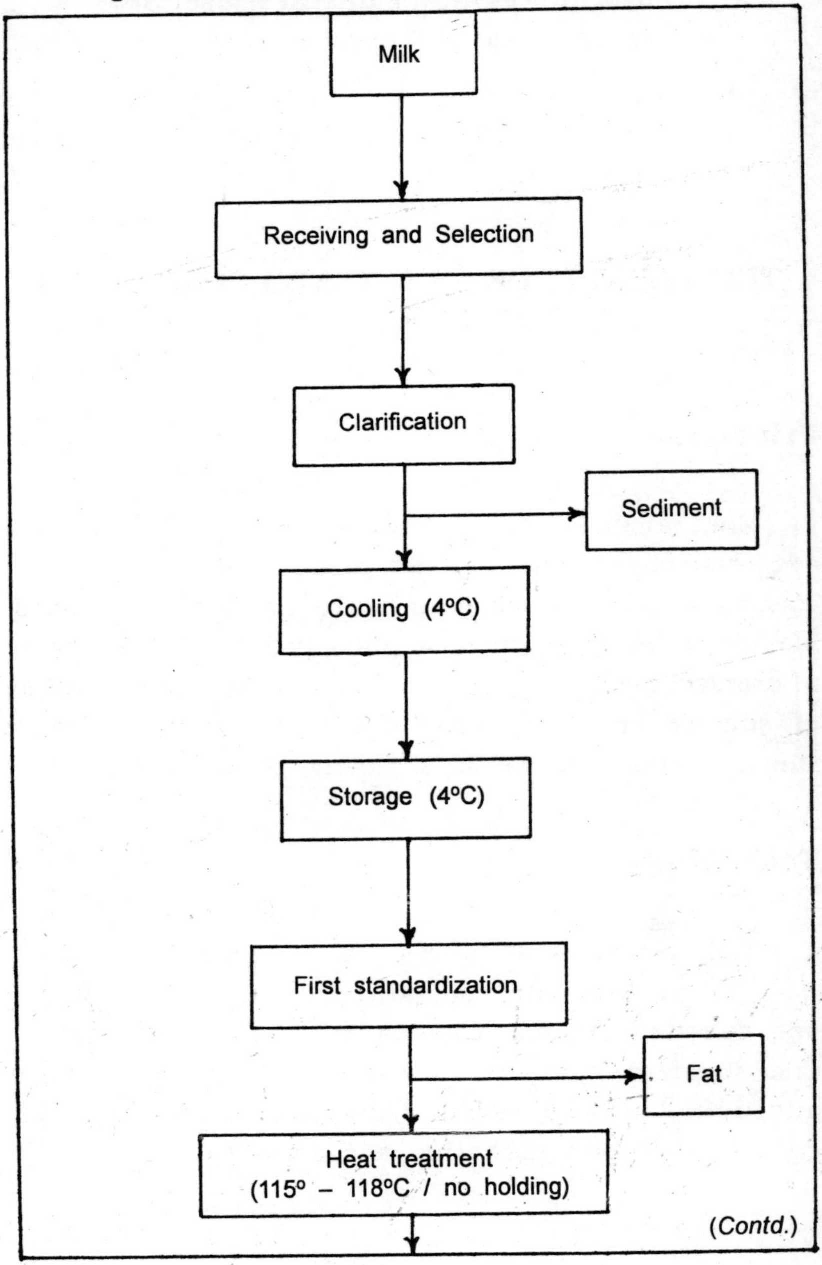

(Contd.)

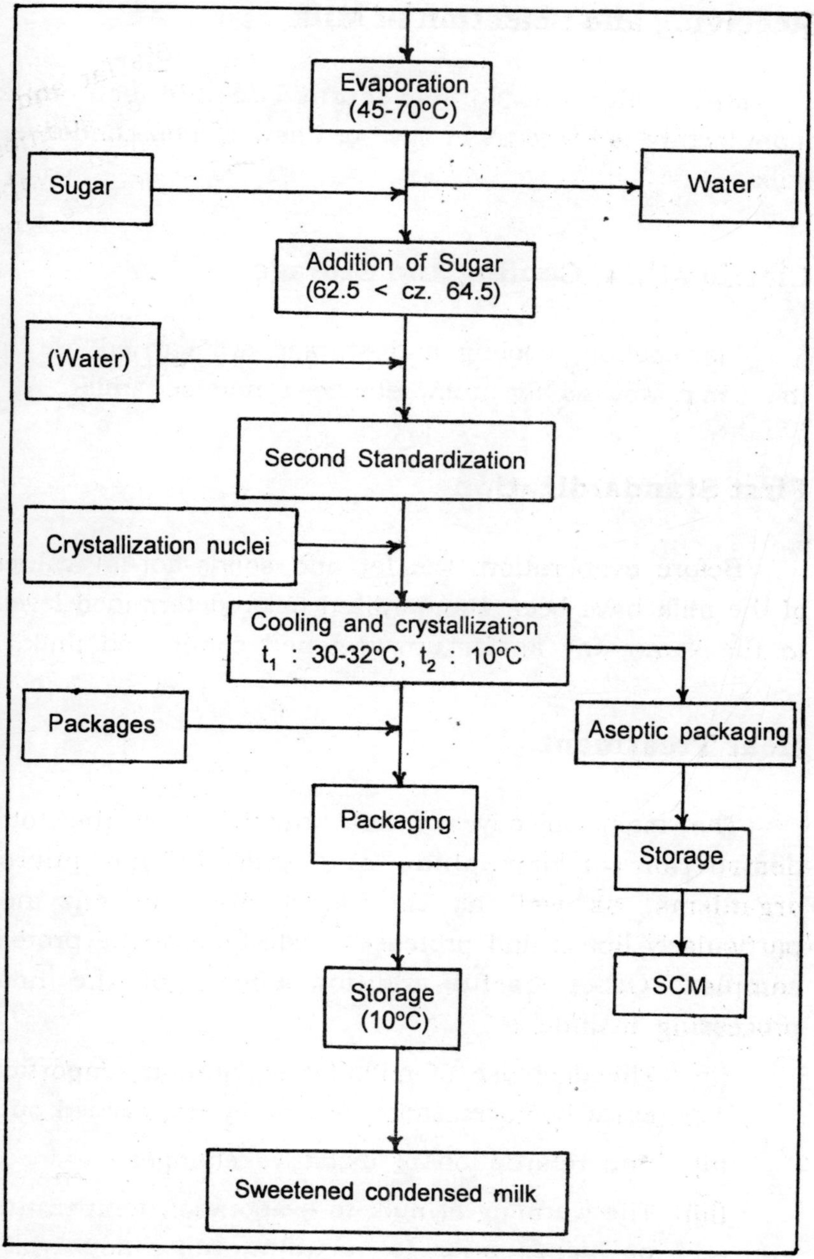
Evaporation
(45-70ºC)
Sugar
Water
Addition of Sugar
(62.5 < cz. 64.5)
(Water)
Second Standardization
Crystallization nuclei
Cooling and crystallization
t1 : 30-32ºC, t2 : 10ºC
Packages
Aseptic packaging
Packaging
Storage
SCM
Storage
(10ºC)
Sweetened condensed milk

Receiving and Selection of Milk

Milk selection for sweetened condensed milk manufacture is same as in case of unsweetened condensed milk.

Clarification, Cooling and Storage

Clarification, cooling and storage are carried out in the same way as for unsweetened condensed milk.

First Standardization

Before evaporation, the fat and solids-not-fat values of the milk have been standardized to predetermined levels in the same way as for unsweetened condensed milk.

Heat Treatment

The main objective of heat treatment is the total destruction of osmophilic and thermophilic micro-organisms, as well as the deactivation of enzymes particularly lipase and proteases and stabilize the protein complex. Other useful addition effects of the heat processing include :

(i) The decrease of milk fat separation (important since homogenization is usually not carried out).

(ii) The retardation of oxidative changes.

(iii) The warming of milk to evaporation temperature or higher prior to vacuum evaporation which has positive economic and technological effects.

Time and Temperature

Heat treatment is important to the development of product viscosity during storage and is particularly important in the case of SCM. A gel can form if the heat treatment is too severe—the milk is usually heat-treated at 82°C for 10 minutes if a product with a relatively high viscosity is required. If a low viscosity product is required, the temperature and time combination should be 116°C/ 30 sec.

Addition of Sugar

Objective

Addition of sugar is the only way to prolong the shelf-life of sweetened condensed milk, since heat sterilization is not used. A sugar content of at least 62.5% in the aqueous phase is required to produce an osmotic pressure higher enough to inhibit the growth of bacteria.

Microbiological Quality of Sugar

Sugar normally does not contain micro-organisms but the precautions are taken to avoid any contamination. It should not contain osmophilic yeasts, mold spores and bacteria producing acid and gas. The standards for sugar used in food preservation industry are given in Table 2.7.

Table 2.7

The Standards of Sugar Used in Food Preservation Industry

Parameter	Specification
Total colony count/g (maxi)	100
Yeasts/g	nil
Molds/g	10
Total thermophilic spores/10 g(maxi)	150
Flat Sour Spores/10 g	To satisfy the test
Thermophilic anaerobes producing gas	Nil
Sulphide spoilage organisms	Nil

Sugar should be stored in a dry place (with less humidity), free from dust, insects and rodent contamination. When sugar is exposed to high humidity in an unsanitary room, it may become moldy. Secondly, sugar should be handled carefully to avoid contamination.

Method of Sugar Addition

Two methods are used :

- Addition of dry sugar before heat treatment.
- Addition of sugar syrup in the evaporator.

Stages of Sugar Addition

The stage at which the sugar is added affects the viscosity of the end product. The early addition of sugar can cause the product to become too viscous during storage.

The sugar is dissolved in water at 95°. Before addition into milk, the solution is heated using a high-pasteurization method to destroy microorganism. The sugar syrup is incorporated late in the condensing process. Alternatively, the sugar may be directly added to the pre-warmed milk before condensing in vacuum pan. The sugar results in plasmolyis of organisms present in milk drawn to the vacuum pan. However, sugar should not be added to milk before pre-heat treatment because it increases the resistance of organisms.

Amount of Sugar Added

The amount of sugar to be added has to be such that its concentration in aqueous phase of the final product ranges from 62.5 to 64.5% (Hunziker, 1949). This sugar ratio (sugar-in-water concentration) not only ensures proper protection against microbial growths but also prevent sugar crystallization.

Formulae

(i) Sugar ratio

$$= \frac{\%\ \text{Sugar in condensed milk}}{100 - \%\ \text{total milk solids in condensed milk}} \times 100$$

(ii) Sugar ratio

$$= \frac{\%\ \text{Sugar content in condensed milk}}{\%\ \text{Sugar content in condensed milk} + \text{water content in condensed milk}} \times 100$$

(iii) % Sugar in condensed milk

$$= \frac{(100 - \%\ \text{total milk solids in condensed milk}) \times \text{Sugar ratio}}{100} \times 100$$

Evaporation

In the condensing operation, milk and sugar mixture is subjected to vacuum evaporation in a vacuum pan (vacuum 25 inch) at approximately 57.2°C at which it boils. The time taken for condensing may be 3-4 hrs. The dry matter content is checked indirectly by determining the density of the concentrate.

Second Standardization

Final standardization is carried out after evaporation and sugar addition. During this stage, total solids, sugar and fat content are controlled. On the basis of the data obtained, the ratio of these components to total milk solids as well as the total dry matter of sweetened condensed milk are adjusted. Some manufacturers prefer to slightly over-condense and then adjust the composition by adding sterile water or under-condensed milk. Such a process is prone to post-contamination and should be very carefully accomplished. A technologically better procedure is to adjust the composition of standardized milk and also to suitably alter the finishing of the batch mixes.

Cooling and Crystallization

Sweetened condensed milk must be cooled after evaporation. This is the most critical and important stage in the whole process. The water in the condensed milk can only hold half of the quantity of lactose in solution. The remaining half will, therefore, be precipitated in the form of crystals, if the surplus lactose is allowed to precipitate freely, the sugar crystals, will be large and the product will be gritty and unsuitable for many applications. It is consequently preferable to control the crystallization of lactose, so that very small crystals are obtained. The largest crystal size permitted in first-grade milk is 10 mm. These crystals will remain dispersed in the milk under normal storage temperature 15-25°C, and are not felt on the tongue.

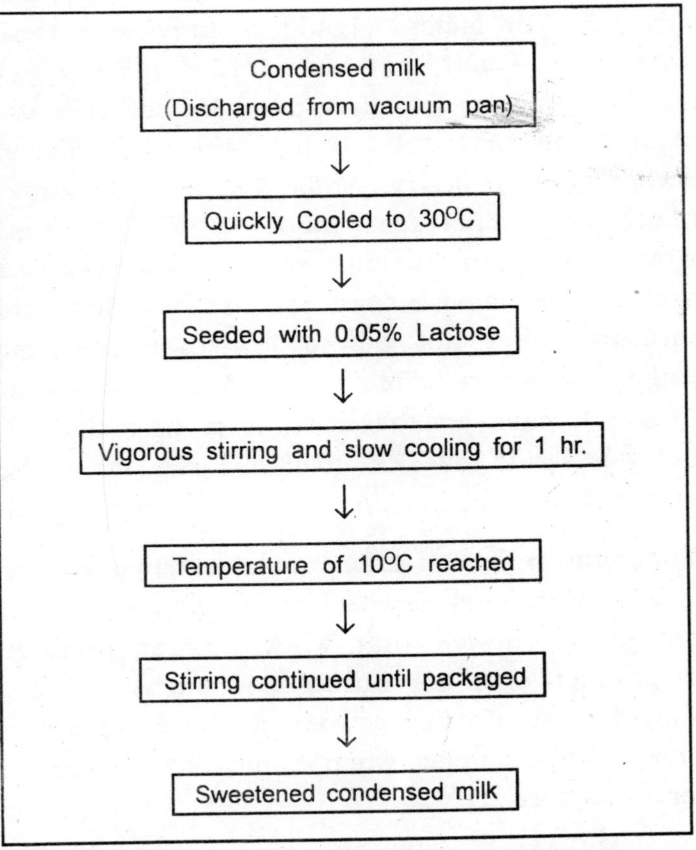

Fig. 2.4 Flow Diagram of Cooling and Crystallization of Sweetened Condensed Milk

The required crystallization is accomplished by cooling the mixture rapidly under vigorous agitation without air being entrapped, and very often by flash cooling (Fig. 2.4).

Seeding is the process of providing nuclei for the development of lactose crystals. Seeding can be done either by adding sterilized lactose or good quality condensed milk from a previous batch.

It is necessary that the seed should be bacteriologically safe. The lactose should be previously sterilized and total count should not exceed 10 to 15 per g. Seed crystals in the form of finely ground lactose crystals, are added at a rate of about 0.05% of the total mix, either as powder or as a slurry, when the milk has cooled to crystallization temperature (about 30°C). This is the temperature at which the sugar solution is supersaturated, so that the seed lactose does not dissolve. However, the temperature must not be so low that spontaneous nucleation can occur before the seed crystals are mixed in. To avoid mass crystallization it is best to cool the product quickly to 39-32°C (which varies with the dry matter concentration) inoculate and cool further to 10°C.

Comparative investigations of sweetened condensed milk seeded with whey powder versus lactose in the same concentrations showed that whey powder produced a better product shelf-life during 3 months storage. In addition, the use of whey powder in this process is more economical and simpler when compared to skim milk powder or lactose.

Equipment Used for Cooling and Crystallization

Cooling with crystallization is accomplished by using double-wall tank-crystallizers, fluid-flow continual coolers, or vacuum crystallizers.

1. Double wall crystallizers resemble double jacketed containers with an especially strong mixer. The water for cooling flows between the walls, while scrapes remove the product from the contact surface.

2. Fluid-flow continual coolers are mainly of tubular type (internal tubular) to accommodate an increased product viscosity. These two cooling methods could be combined, with the lactose crystals inoculated applied in between.

3. Vacuum crystallizers are units, similar to vacuum evaporators, in which water is evaporated at a reduced pressure of about 9.3-9.6 Kpa, cooling the product and causing lactose crystallization. Crystallization nuclei are dosed automatically.

   The advantages of this equipment for lactose crystallization are :

   - Increase in crystallization rate
   - High capacity
   - Decreased reinfection possibility
   - No air occlusion
   - Automatic deaeration, thus, decreasing oxidation chances of milk fat.

Packaging and Storage

Retail use

Retail sweetened condensed milk is mostly packed in tin cans. Can used for filling sweetened condensed milk should be cleaned and sterilized before filling, as no sterilization takes place after canning. Lids should also be sterilized properly before use. The cans must be filled as full as possible to minimize head space otherwise aerobic organisms, if any, present, may grow and cause defect in the product. Now-a-days, it is also possible to pack SCM in aseptic paper board packages, tubes, various plastic forms and other containers.

Industrial supply

The product is also packed in beech-wood barrels, metal barrels, metal cylindric drums and other containers, holding about 300 kg., for supply to large-scale users.

Storage

The storage of sweetened condensed milk is accomplished under 10°C. According to Radaveva *et al.* (1982) the sweetened condensed milk could be stored for many years at 10°C without any compositional changes. The only condition was that during processing, the milk be optimally heated and homogenized. Coffee extract, if present in this product, protects it from fat oxidation for a few years of storage.

FREEZING OF SWEETENED CONDENSED MILK

The average freezing point of sweetened condensed milk is –15°C. At a storage temperature below normal and with the progress of time, the intensity of lactose crystallization increases leading to a coarse texture and possible sedimentation even at temperatures above freezing point. In some cases, even sucrose may crystallize out. With the freezing of SCM, these effects may be accelerated and in extreme cases desterilization of fat and flocculence may also appear depending upon the degree of crystallization and storage priod (Webb *et al.*, 1978).

Judging and Grading

High-grade condensed milk should have clear, pleasant aroma, a pronounced sweet taste, smooth and uniform body and texture, uniform light colour. The score card prepared by American Dairy Science Association is given in Table 2.8.

Table 2.8

A Score Card for Judging and Grading of SCM

Items	Perfect Score
Flavour and odour	30
Body and texture	25
Colour	5
Fat content	10
Milk solids	10
Bacteria	10
Sugar	10

Problem 1

15,000 kg of milk having 6.5% fat and 9% S.N.F. is standardized for SCM manufacture by separation of 60% fat cream. If desired composition of SCM is 9.5% fat and 22% S.N.F. and 40% sugar, then calculate the

(i) amount of cream to be separated for SCM

(ii) amount of SCM

(iii) amount of sugar

(iv) ratio of concentration

(v) sugar ratio.

Solution

(i) Suppose, x kg cream to be separated for SCM. Now, calculate total amount of fat in 15,000 kg. of milk.

100 kg milk contains 6.5 kg fat.

$\therefore$ 15,000 kg. milk contains $\frac{6.5 \times 15,000}{100}$

$= 975$ kg. fat.

100 kg milk contains 9.0 kg SNF

$\therefore$ 15,000 kg. milk contains $\frac{9 \times 15,000}{100}$

$= 1350$ kg SNF

100 kg cream contains 60 kg fat.

$\therefore$ x kg. cream contains $\frac{60}{100} \times x = 0.6\,x$ kg. fat.

Now, percentage of SNF in cream

$$= \frac{100 - \%\text{ fat in cream}}{100 - \%\text{ fat in milk}} \times \%\text{ SNF in milk}$$

$$= \frac{100-60}{100-6.5} \times 9 = \frac{40 \times 9}{93.5} = 3.85\%\text{ SNF}$$

100 kg cream contains 3.85 kg. SNF

$\therefore$ x kg. cream contains $\frac{3.85}{100} \times x = 0.0385x$ kg SNF

Therefore, the amount of standardized milk

$= 15{,}000 - x$ kg.

Amount of fat in standardized milk $= 975 - 0.6x$ kg.

Amount of SNF in standardized milk

$= 1350 - 0.0385x$ kg.

$\therefore$ Fat : SNF : : Fat : SNF in SCM

9.5 : 22 : : 975 − 0.6x : 1350 − 0.0385x

Or $\frac{9.5}{22} = \frac{975 - 0.6x}{1350 - 0.0385x}$

or $9.5\,(1350 - 0.0385x) = 22\,(975 - 0.6x)$

or $12825 - 0.36575x = 21450 - 13.2x$

or $13.2x - 0.36575x = 21450 - 12825$

or $12.83425x = 8625$

or $x = \frac{8625}{12.83425} = 672.03$

∴ 672.03 kg cream to be separated for SCM.

Amount of cream to be separated for SCM

= 672.03 kg.

(ii) Amount of standardized milk = (15000 – x) kg

= 15000 – 672.03 = 14327.97 kg.

∴ Amount of fat in standardized milk = 975 – 0.6x

= 975 – 0.6 × 672.03

= 975 – 403.218 = 571.782 kg.

Amount of SNF in standardized milk

= 1350 – 0.0385x

= 1350 – 0.0385 × 672.03

= 1350 – 25.873155

= 1350 – 25.87

= 1324.13 kg.

∴ Total T.M.S. = Amount of fat in standardized milk + Amount of SNF in standardized milk

= 571.781 + 1324.13

= 1895.911 kg

9.5 kg fat available from 100 kg. SCM.

∴ 571.781 kg. fat available from = $\frac{100}{9.5} \times 571.781$

= 6018.75 kg.

Amount of SCM = 6018.75 kg.

(iii) 100 kg SCM contains 40 kg. sugar

$\therefore$ 6018.75 kg. SCM contain $\frac{40}{100} \times 6018.75$ kg. sugar

$= 2407.5$ kg.

Amount of Sugar = 2407.5 kg.

(iv) Total milk solid in standard milk = 1895.91 kg.

$\therefore$ 14327.97 kg. standard milk contains 1895.91 kg. TMS

$\therefore$ 100 kg. standard milk contains $\frac{1895.91}{14327.97} \times 100$

$= 13.23$ kg

$\therefore$ Percentage of total milk solid in standardized milk $= 13.23\%$

$$\therefore \text{Ratio of concentration} = \frac{\%\ \text{TMS in condensed milk}}{\%\ \text{TMS in milk}}$$

$$= \frac{31.50}{13.23} = 2.38$$

Ratio of concentration = 2.38

$$\text{(v) Sugar ratio} = \frac{\%\ \text{Sugar}}{100 - \%\ \text{TMS}} \times 100$$

$$= \frac{40}{100 - 31.5} \times 100 = 58.39$$

$\therefore$ **Sugar ratio = 58.39%**

Problem 2

12,000 kg. of milk having 5.5% fat and 9% SNF is standardized for SCM manufacture by addition of skim milk having 0.2% fat, 9% SNF. If desired composition of SCM is 9.5% fat and 23% MSNF and 43% sugar, then calculate :

(i) Amount of skim milk

(ii) Amount of standardized milk

(iii) Amount of SCM

(iv) Amount of sugar

(v) Ratio of concentration

(vi) Sugar ratio.

Solution

Suppose, x kg. skim milk is added for SCM manufacturing,

∴ Total amount of fat in 12000 kg. of milk

100 kg. milk contains 5.5 kg. fat

$\therefore$ 1 kg. milk contains $\frac{5.5}{100}$ kg. fat

$\therefore$ 12,000 kg. milk contains $\frac{5.5 \times 12{,}000}{100} = 660$ kg. fat

100 kg milk contains 9 kg. SNF

$\therefore$ 1 kg milk contains $\frac{9}{100}$ kg. SNF

$\therefore$ 12,000 kg milk contains $\frac{9}{100} \times 12,000$

$= 1,080$ kg. SNF

100 kg. skim milk contains 0.2 kg. fat

$\therefore$ 1 kg. skim milk contains $\frac{0.2}{100}$ kg. fat

$\therefore$ x kg. skim milk contains $\frac{0.2}{100} \times x$

$= 0.002x$ kg. fat.

100 kg. skim milk contains 9 kg. SNF

$\therefore$ 1 kg. skim milk contains $\frac{9}{100}$ kg SNF

$\therefore$ x kg. skim milk contains $\frac{9}{100} \times x = 0.09\ x$ kg. SNF

The amount of standardized milk = (12,000 + x) kg.

$\therefore$ Amount of fat in standardized milk

$= 660 + 0.002x$ kg. fat

$\therefore$ Amount of SNF in standardized milk

$= 1080 + 0.09x$ kg. SNF

$\therefore$ Fat : SNF :: Fat : SNF. (in SCM).

$\therefore$ $9.5 : 23 :: (660 + 0.002x) : (1080 + 0.09x)$

$$\frac{9.5}{23} = \frac{660 + 0.002x}{1080 + 0.09x}$$

$\therefore$ $9.5\ (1080 + 0.09x) = 23\ (660 + 0.002x)$

or $10260 + 0.855x = 15180 + 0.046x$

or $0.855x - 0.046x = 15180 - 10260$

$$0.809x = 4920$$

Or $$x = \frac{4920}{0.809} = 6081.58 \text{ kg.}$$

$$x = 6081.58 \text{ kg.}$$

∴ **Amount of skim milk 6081.58 kg.**

(ii) Amount of Standardized milk = 12000 + x

= 12000 + 6081.58

= **18081.58 kg.**

Amount of fat in standardized milk

= 660 + 0.002x

= 600 + 0.002 × 6081.58 = 672.16 kg.

Amount of SNF in standardized milk

1080 + 0.09x = 1080 + 0.09 × 6081.58 = 1627.34 kg.

Total T.M.S. = (672.16 + 1627.34) kg.

= 2299.5 kg.

(iii) ∴ 9.5 kg. fat contains in 100 kg. SCM.

1 kg. fat contains in 100 kg. SCM = $\frac{100}{9.5}$ Kg. SCM

672.16 kg. fat contains in 100 kg SCM

$$= \frac{100}{9.5} \times 672.16 = 7075.37 \text{ kg.}$$

Amount of SCM = 7075.37 kg.

(iv) 100 kg. SCM contain 43 kg. sugar.

1 kg. SCM contains $\frac{43}{100}$ kg. sugar

7075.37 kg. SCM contain $\frac{43}{100} \times 7075.37 = 3042.41$ kg.

Amount of sugar = 3042.41 kg.

(v) T.M.S. in standardized milk = 2299.5 kg.

18081.58 kg. std. milk contains = 2299.5 kg.

1 kg. std. milk contains $= \frac{2299.5}{18081.58}$

100 kg. std. milk contains $= \frac{2299.5}{18081.58} \times 100$.

$= 12.72$ kg.

∴ **% of TMS in standardized milk = 12.72**

∴ Ratio of concentration $= \frac{\text{\% TMS in condensed milk}}{\text{\% TMS in milk}}$

$= \frac{32.5}{12.69} = 2.56$

∴ **Ratio of concentration = 2.56**

Sugar ratio $= \frac{\text{\% Sugar}}{100 - \text{TMS\%}} \times 100$

$= \frac{43}{100 - 32.5} \times 100 = 63.7\%$

∴ **Sugar ratio = 63.7%**

Problem 3

8000 kgs. of milk, testing 6% fat and 9.5% SNF is standardized by addition of skim milk of 0.5% fat and 8.5% SNF. If desired composition of SCM 9% fat and 22% SNF and 42% sugar, then calculate :

(i) Amount of skim milk

(ii) Amount of standardized milk

(iii) Amount of SCM

(iv) Amount of sugar

(v) Ratio of concentration

(vi) Sugar ratio.

Solution

Suppose x kg. of skim milk is added for SCM

100 kg. of milk contains 6 kg fat.

$\therefore$ 1 kg. of milk contains $\frac{6}{100}$ kg fat.

$\therefore$ 8000 kg. of milk contains $\frac{6}{100} \times 8000$

$= 480$ kg. fat

100 kg. of milk contains 9.5 kg. SNF

$\therefore$ 1 kg. of milk contains $\frac{9.5}{100}$ kg. SNF

8000 kg. of milk contains $\frac{9.5}{100} \times 8,000$

$= 760.0$ kg. SNF

100 kg. skim milk contains 0.5 kg. fat

$\therefore$ 1 kg. skim milk contains $\frac{0.5}{100}$ kg. fat

$$\frac{0.5}{100} \times x = 0.005x \text{ kg. fat}$$

100 kg. of skim milk contains 8.5 kg. SNF

1 kg. of skim milk contains $\frac{8.5}{100}$ kg. SNF

x kg. of skim milk contains $\frac{8.5}{100} \times x$

$= 0.85x$ kg.

The amount of standardized milk = (18000 + x) kg.

Amount of fat in standardized milk

$= (480 + 0.005x)$ kg.

Amount of SNF in standardized milk

$(760 + 0.085x)$ kg.

$\therefore$ Fat : SNF :: Fat : SNF (in SCM)

9 : 22 :: (480 + 0.005x) : (760 + 0.085x)

or $$\frac{9}{22} = \frac{480 + 0.005x}{760 + 0.085x}$$

or 9 (760 + 0.085x) = 22 (480 + 0.005x)

or 6840 + 0.765x = 10560 + 0.11x

or $x = \frac{3720}{0.655}$

x = 5679.39 kg.

Amount of Skim milk 5679.39 kg.

Amount of standardized milk = (8,000 + x) kgs.

= 8,000 + 5679.39

= 13679.39 kgs.

Amount of standardized milk = 13679.39 kgs.

Amount of fat in standardized milk

= (480 + 0.005x)

= 480 + 0.005 × 5679.39 = 508.4 kg.

Amount of SNF in standardized milk

= 760 + 0.085x

= 760 + 0.085 × 5679.39 = 1242.75 kg.

Amount of TMS in standardized milk

= 508.4 + 1242.75 = 1751.15 kg.

9 kg. fat contains 100 kg. SCM

1 kg. fat contains $\frac{100}{9}$ kg. SCM

508.4 kg. fat contains $\frac{100}{9} \times 508.4 = 5648.89$ kg.

Amount of SCM = 5648.89 kg.

100 kg. SCM contains 42 kg. sugar.

∴ 1 kg. SCM contains $\frac{42}{100}$ kg. sugar.

5648.89 kg. SCM contains $\frac{42}{100} \times 5648.89$

$= 2372.53$ kg.

$\therefore$ **Amount of Sugar = 2372.53 kg.**

13679.39 kg. standardized milk contain 1751.15 kg. TMS

$\therefore$ 1 kg. standardized milk contains $= \frac{1751.15}{13679.39}$

$\therefore$ 100 kg. standardized milk contains $= \frac{1751.15}{13679.39} \times 100$

$= 12.8$ kg.

$\therefore$ % of TMS in standardized milk = 12.8%

$$\text{Ratio of concentration} = \frac{\text{\% TMS in SCM}}{\text{\%TMS in milk}}$$

$$= \frac{31}{12-8} = 2.42$$

Ratio of concentration = 2.42

$$\text{Sugar ratio} = \frac{\text{\% Sugar}}{100 - \text{\%TMS}} \times 100$$

$$= \frac{42}{100-31} \times 100 = 60.87$$

Sugar ratio = 60.87%

DISCUSSION QUESTIONS

1. Define evaporated milk. Explain its chemical composition and nutritive value.
2. Explain the manufacturing technology of evaporated milk.
3. What is the basic difference between evaporated and sweetened condensed milk ? Explain the role of sugar in sweetened condensed milk and how is the amount of sugar calculated ?
4. Explain the manufacturing process of sweetened condensed milk.
5. What do you understand by cooling and crystallization in SCM ?

3

Evaporators

The problem of water activity in the dairy products and resultant spoilage became the greatest worry for dairy industry, till evaporators came in as a major rescue. With the help of evaporators, milk is partially condensed by the removal of water. Reduction in water volume facilitates milk to transport in an economical way due to lesser weight. Modern Technological advances helped to introduce various types of evaporators that made condensing rather effective, concentration of liquid by evaporation under vacuum was introduced in 1913. The process was based on a British patent by E.C. Howard, which covered a steam-heated double-bottomed vacuum pan with condenser and vacuum pump.

Principles of Evaporation

Evaporation is used to boil off a portion of water from milk. Milk evaporator basically consists of a heat exchanger to supply sensible heat and latent heat of vaporization to milk, a separator in which the vapour is separated from the concentrated milk and a condenser to effect condensation of the vapour and its removal from the system (Fig. 3.1).

Selection of Evaporator

Milk properties greatly influence the selection of evaporators and its operation. These are viscosity, temperature sensitivity, foaming, fouling and corrosion.

(i) *Viscosity*

Higher viscosities result in lower heat transfer coefficient and reduced circulation rates. The viscosity of milk, undergoing evaporation, increases with concentration, which, therefore, results in decrease in heat transfer rate and, hence, capacity.

(ii) *Temperature sensitivity*

Milk is sensitive to heat. Therefore, to reduce the risk of heat damage during evaporation, low boiling

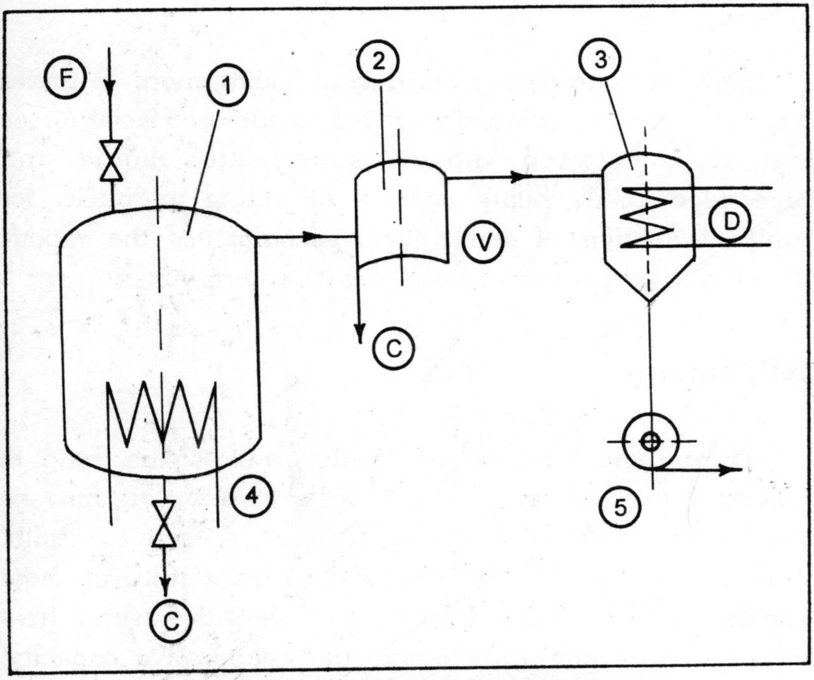

Fig. 3.1 Principles of evaporation

1. Evaporator, 2. Separator, 3. Condenser, 4. Heater, 5. Vacuum pump and condensate pump,

F = Feed, C = Concentrate. D = Cooling. V = Vapour.

temperatures and shorter residence times should be used. Low boiling temperatures are obtained by operating the evaporator at low pressure, more specifically under vacuum. Satisfactory temperature differences can then be achieved with heating medium (steam) at correspondingly lower temperature.

(iii) *Foaming*

Milk tends to form foam during boiling more so under vacuum. Foaming is seemly related to the interfacial forces that arise between vapour, superheated liquid and suspended solids being capable of acting as nuclei for bubble formation. A stable foam accompanies the vapour out of the evaporator causing heavy entertainment.

(iv) *Fouling*

During the evaporation of milk, milk solids, tend to adhere to the heat transfer surfaces. This fouling may be due to adhesion of suspended solids present in milk. More usually, the scale is deposited from milk on heat transfer surface during evaporation lowers the overall heat transfer coefficient and, hence, the evaporative capacity. Another problem due to fouling is more frequent and more difficult cleaning.

(v) *Corrosion characteristics*

Milk partially dissolves copper and its allies, tin, aluminium etc. Stainless steel is not dissolved in milk and for this and for many reasons, evaporator and other pieces of equipments coming in contact with milk are made of stainless steel.

EVAPORATION IN DAIRY INDUSTRY

In the dairy industry, evaporation is used to concentrate whole milk, skim milk, whey, whey protein concentrate and permeate from membrane filtration modules.

EVAPORATORS AND ANCILLARY EQUIPMENTS

Evaporators

Tubular and plate type evaporators are the basic types of evaporators used in recent years.

Vacuum Pan Evaporator

It consists of evaporating vessel with heating surface, vapour space, condenser, vacuum pan and other accessories. The heating surface consists of 2 or 3 short coils. (Fig. 3.2). The portion of the pan above heating surface is large vapour space where thermometer, vacuum break, milk inlet pipe, observation and illumination glasses are provided. This plant is suited for batch operation. It was superseded by tubular evaporator.

Tubular Evaporator

There are different types of tubular evaporators (Fig. 3.3). They are classified as :

(a) Position of tubes : Horizontal and vertical.

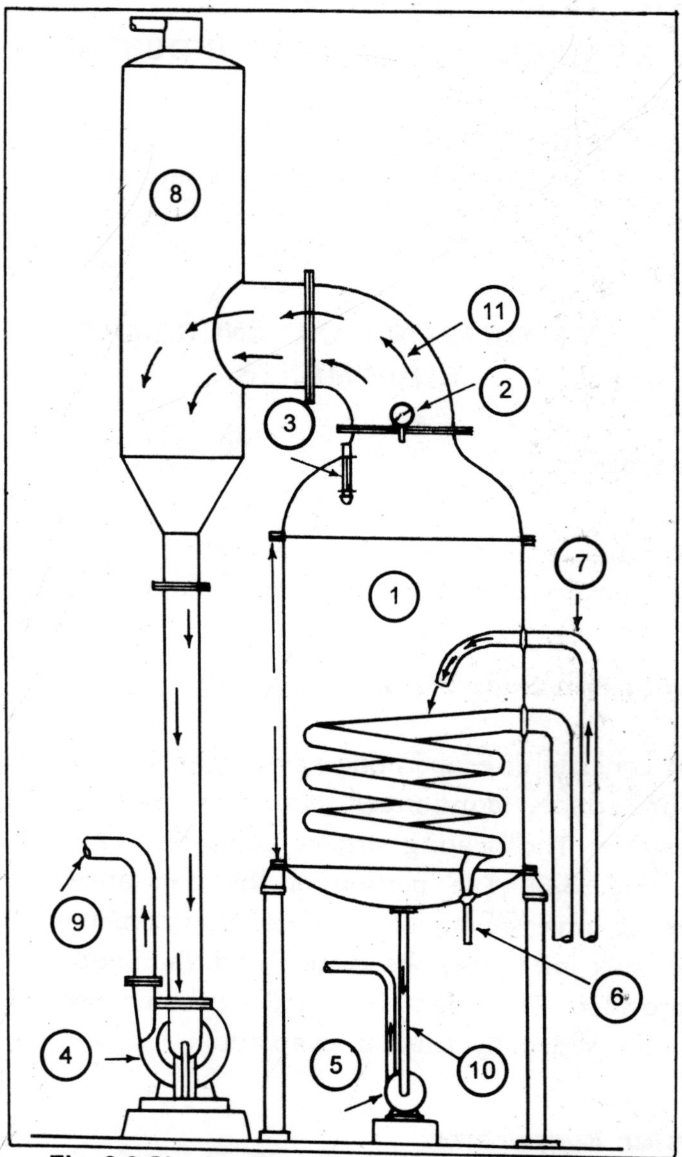

Fig. 3.2 Single effect vacuum Pan Evaporator
(***Courtesy*** : Evaporated Milk Association)

1. Vacuum pan, 2. Vacuum gauge, 3. Thermometer, 4. Water pump, 5. Evaporated milk pump, 6. Steam inlet, 7. Inlet for milk. 8. Water vapour condenser, 9. Water remover, 10. Outlet for freshly evaporated milk, 11. Water vapour from milk.

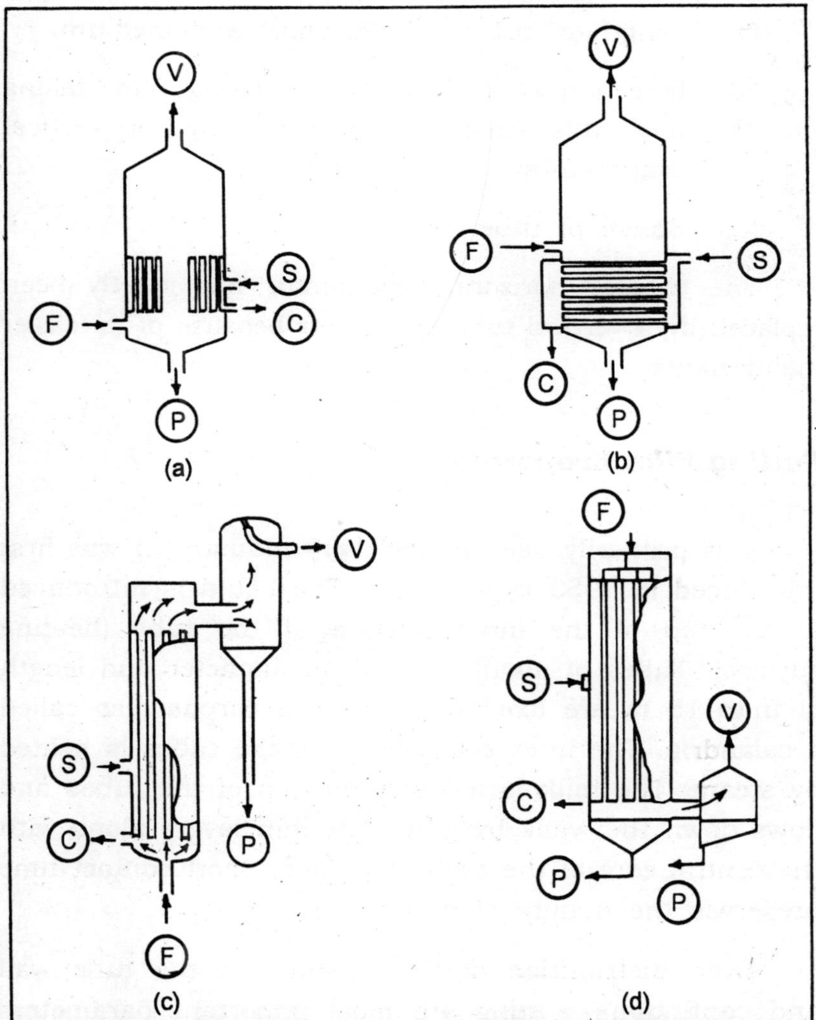

Fig. 3.3 Different types of tube evaporator constructions

(a) Vertical short tube evaporator,

(b) Horizontal tube evaporator,

(c) Vertical long tube evaporator with rising film,

(d) vertical falling film evaporator (long tube)

F= Feed, S = Steam, V = vapour, C = Condensate, P = Product

(b) Length of tubes : long, short and medium.

(c) Direction of product flow : rising film, falling film (this classification is mainly for vertical evaporators)

(d) Shape of tube.

The tubular horizontal evaporator has mostly been replaced by a vertical tube evaporator because of its easier maintenance.

Falling Film Evaporator

It is primarily used in the dairy industry. It was first introduced in 1953 in Germany. The liquid is introduced at the top of the inner surface of the tube (heating surface). Tubes of about 3-5 cm. in diameter and length of upto 15 m are fixed together in a corpus also called a calandria. The inter space between the tubes is heated by steam. The milk is fed into the top of the tubes and flows down the walls by gravity at thin layers along with the central core of the evolved vapour. Short contact time preserves the quality of milk.

Even distribution of the product on the tube wall and continuous wetting are most important parameters for successful operation of the evaporator. The product is preheated at a slight pressure somewhat above the temperature of generation in the evaporator. By creating a vacuum, the temperature in the evaporator will be below 100°C. By means of a nozzle shaped device, the product is distributed over the spreader plate. Since the milk is somewhat super-heated, it expands as soon as it leaves

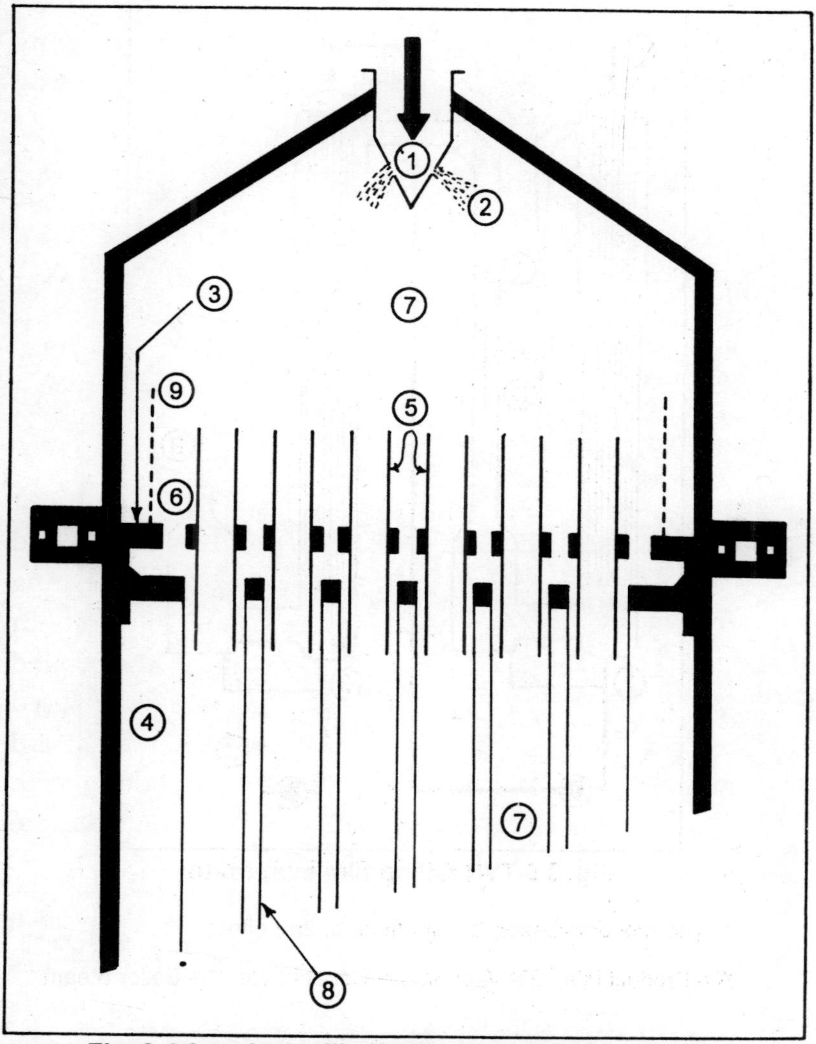

Fig. 3.4 A typical milk distribution system in Falling Film Evaporator

1. Spray nozzle, 2. Feed, 3. Distributor plate, 4. Steam for heating, 5. Coaxial tube, 6. Opening, 7. Vapour, 8. Evaporation tube, 9. Strainer

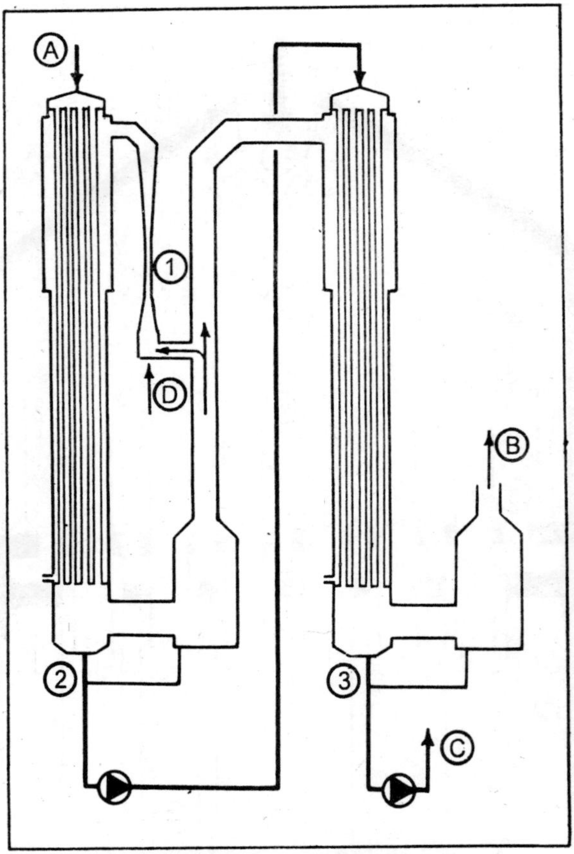

Fig. 3.5 TVR falling film evaporator

1. Thermocompressor, 2. 1st effect, 3. 2nd effect.

A = Product inlet, B = Vapour, C = Concentrate, D = Boiler Steam.

the nozzle, where part of the water is immediately vapourized. The steam, thus, formed press the milk outwards against the inside of the tubes in a thick film which runs down along the walls of the tubes. The water in the film evaporators rapidly flows as the milk passes through the tubes (Figs. 3.4 and 3.5). A steam separator is fitted under the evaporator to separate the steam from the concentrated milk. The holding time in the falling evaporator is very short, about one minute as only a small amount of the product is treated at the same time. The compressor fan draws off the vapour from the collector and compresses it. The compressed vapour is forced into the casing of heater, where it then condenses on the tuber surface of the tubes. The condensate is pumped out and used for pre-heating the feed product.

Multiple-effect evaporators

Multiple-effect evapo-rators are usually used in the dairy industry. The theory is that if two evaporators are connected in series, the second effect can operate at a

Table 3.1

Typical Consumptions per kg. Evaporated Water for Falling Film Tubular Evaporators with Thermal Recompression in the Dairy Industry

Type	Steam
2-effect	0.32 kg.
3-effect	0.25 kg.
4-effect	0.18 kg.
5-effect	0.13 kg.
6-effect	0.11 kg.
7-effect	0.09 kg.
1-effect MVR	0.012 kwh.

Source : Dairy Processing Handbook.

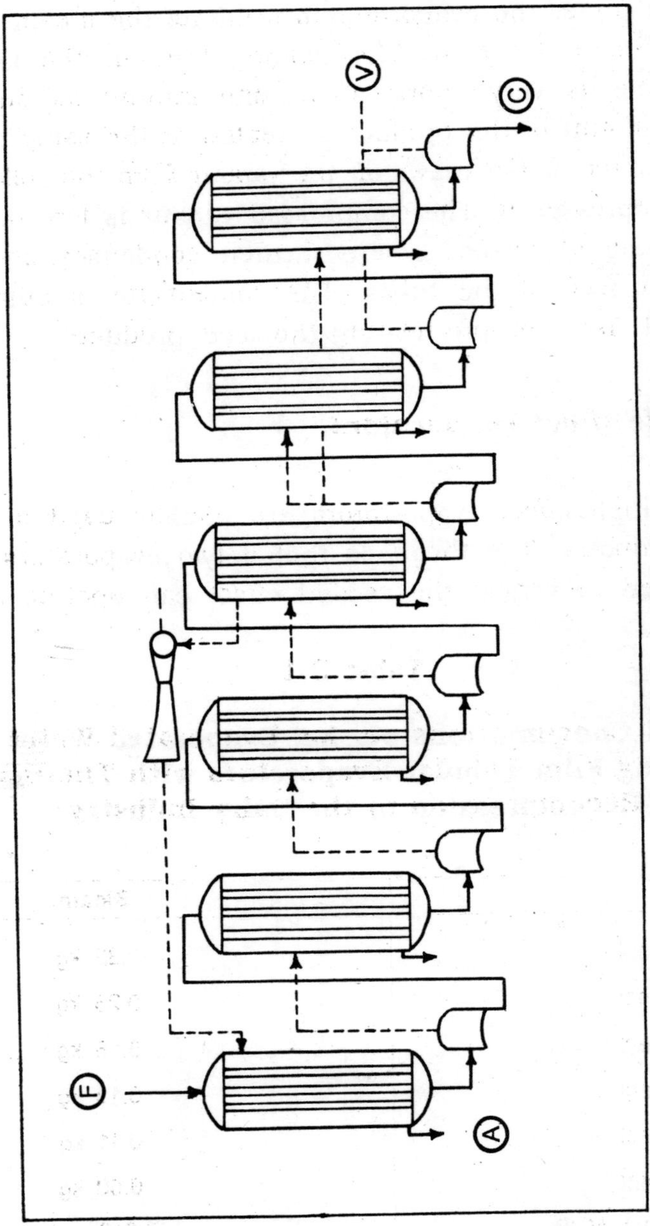

Fig. 3.6 Multiple effect evaporator

F = Feed, A = Condensate, C = Concentrate, V = Vapour.

higher vacuum (and, therefore, at a lower temperature) than the first. The vapour evolved from the product in the first effect can be used as the heating medium for the next effect, which operates at a lower boiling temperature due to the higher vacuum (Fig. 3.6, Table 3.1).

Advantage : It improves steam economy.

Disadvantage : This treatment is not effective for heat sensitive product. The actual operating pressures and temperatures in the condensing plant depend upon the design and operative factor (Table 3.2).

Table 3.2

Operating Pressure and Temperature in Multiple Effect Milk Condensing Plant

Milk/Steam/Water	Temperature	Pressure
Preheated Milk	90°-95°C	—
First effect steam	85°-90°C	0.6-0.7 kg/cm abs.
First effect milk	70-75°C	46-52 cm Hg. Vacuum
Second effect milk	60°-65°C	56-61 cm Hg. Vacuum
Third effect milk	45°-50°C	66-68 cm. Hg vacuum
Condensing Cooling water		30°C —

Thermal Vapour Re-compressor (TVR)

It is improved evaporation technique to minimize total energy consumption. The vapour evolved from the product can be compressed and used as a heating medium. A thermo compressor is used for this purpose (Fig. 3.7).

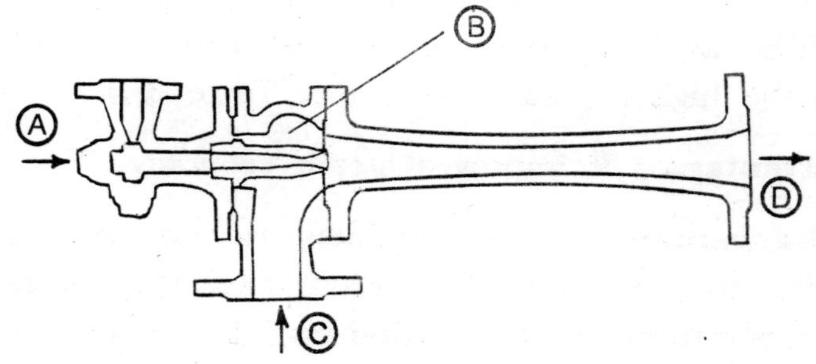

Fig. 3.7 Diagram of Thermo Compressor

A. Boiler Steam, B. Steam nozzle, C. Vapour from separator, D. Recompressed vapour.

Calandria are used as heat exchangers. The tubes are upto 15 m long and 5 cm in diameter. The milk is introduced at the top of the calandria and moves by gravity down the inside surface of the tubes as a thin film aided by the vapour which also moves downwards. The vapour is separated from the milk in a separator placed at the base of the calandria. The vapour is used as the heating medium in the following effect and from the last effect, the vapour goes to the condenser. A part of the vapour is supplied to thermo compressor, to which high pressure steam (6-10 bar) flows.

The compressor uses the high steam pressure to increase the kinetic energy, and the steam is ejected at high speed from the nozzle. This set effect mixes the steam and the vapour from the product and compresses the mixture to a higher pressure. This line steam-vapour mixture is then used to heat the first effect calandria.

The design of a TVR is critical, but the unit is inexpensive, and it saves the steam consumption and requires no maintenance.

Uniform distribution of the milk overall the tube is essential. If the tubes are starved, deposits or scale in the tube will occur quickly.

Mechanical Vapour Recompression (MVR)

It is a newer system and is superior to the conventional ones in areas where electric energy is cheap. Once the evaporator with MVR has been started, very little, or no steam at all, is required for its operation. Mechanical vapour recompression fan (turbo compressor) draws all the vapour out of the evaporator and compresses it before returning it to the heating side of the evaporator. Vapour's heat is used to generate further vapour, so that no heat is rejected into cooling water, making the steam highly efficient. In fact, the only energy required to keep the system in operation is the mechanical energy, which is required to compress the vapour through necessary pressure differential.

As Fig. 3.8 shows, the feed has to be heated to the boiling temperature in the first effect. The heat recovered from the condensate and the concentrate as well as that from the 3% excess steam from the injected condensate should be taken into account. Comparison of TVR and MVR with corresponding heat flow diagram is given in Fig. 3.9.

Advantages of MVR

1. MVR makes it possible to reduce energy requirement.
2. The maximum evaporating temperature can be reduced to such an extent that burning-on of product is minimized.
3. The lowest evaporating temperature can be high enough to minimize bacteriological problems and to lower the viscosity of the product.

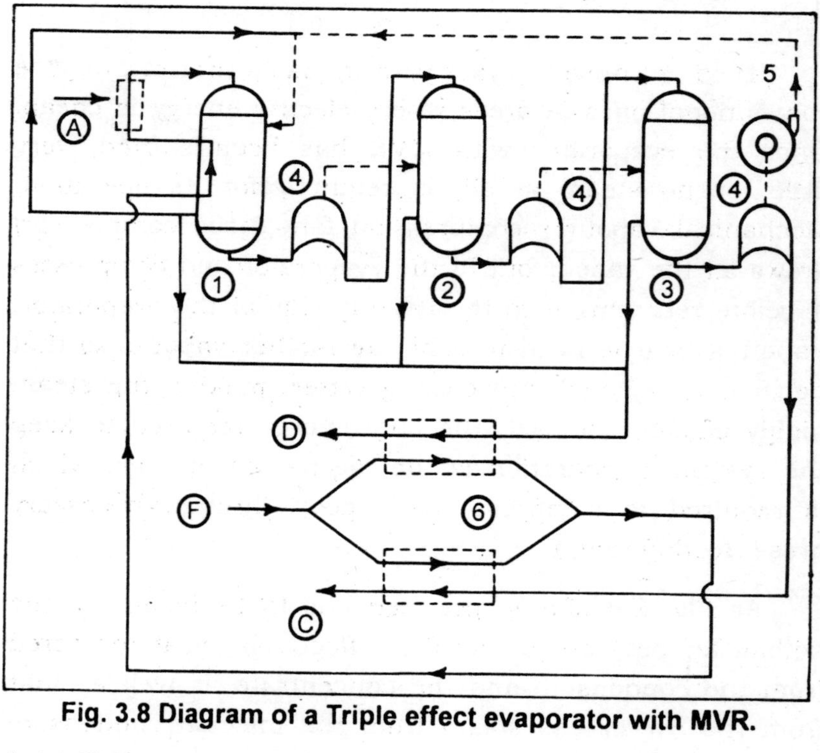

Fig. 3.8 Diagram of a Triple effect evaporator with MVR.

1. 1st effect, 2. 2nd effect, 3. 3rd effect, 4. Separator, 5. MVR, 6. Heat exchanger.

A = Heat, F = Feed, D = Condensate, C = Concentrate.

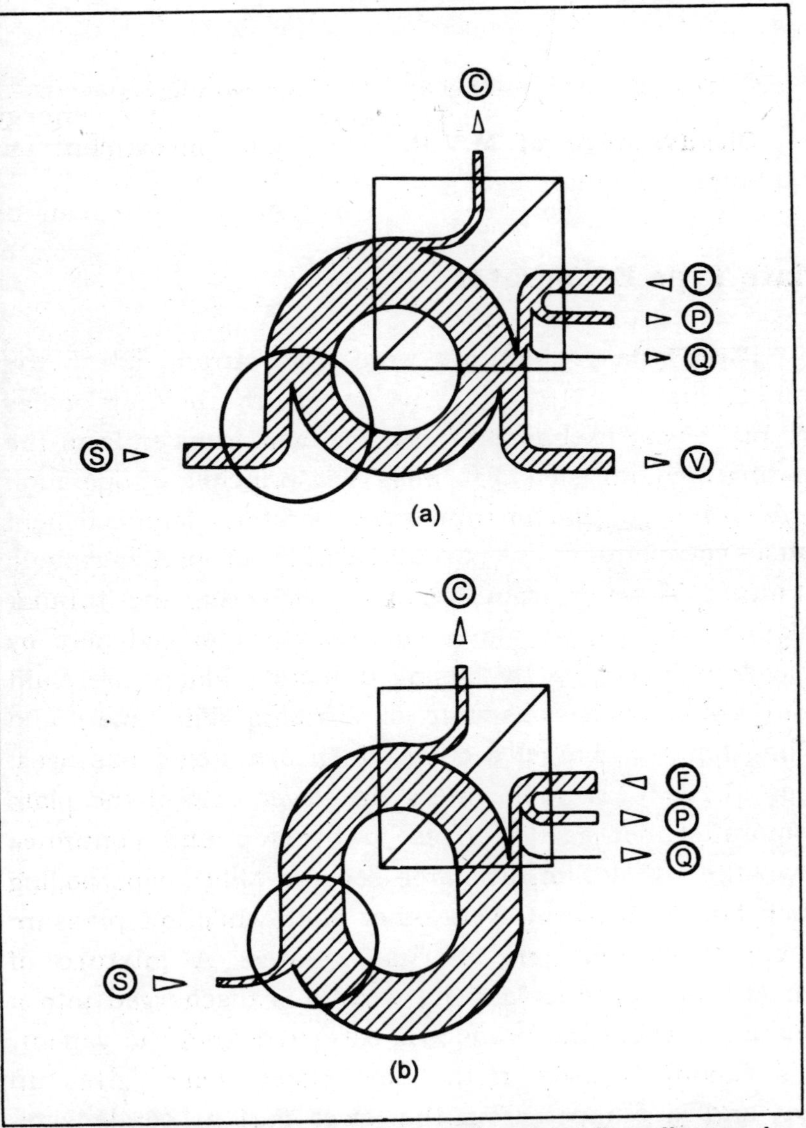

Fig. 3.9 Evaporator with corresponding heat-flow diagram in (a) TVR (b) MVR.

F = Feed, S = Steam, V = Vapour, C = Condensate, Q = Heat loss, P = Product.

4. There is no need for cooling water.

5. Cleaning cycles can be more widely spaced.

Disadvantage of M.V.R.—The initial investment on equipment is quite high.

Plate Type Evaporator

Plate type evaporators were first introduced to the dairy industry in 1957. In this design, the heating bodies are plate heat exchangers need for heat transfer from the heating medium (steam) to milk. The principle of operation is the same as that of tubular evaporators, supplied heat causes evaporation in partial vacuum of predetermined amounts of water from the milk. However, the tubular calandria of conventional evaporators is replaced here by a system of compactly arranged plates. Plates are built upto units, each consisting of a rising film (inlet) and falling film (discharge) section which are steam passages. (Figs. 3.10, 3.11). Milk enters the lower part of the plate evaporator-inlet section rises to the top and continues down the falling film discharge section. Milk keeps boiling when flowing between every other plate while low pressure steam flows between alternate plates. A mixture of concentrated product and its vapour is discharged into a separator, where the product is extracted from the vapour. This vapour is used in the next effect steam, gives up heat and is pumped from the lower part as condensate, which may be either returned to the boiler or used for preheating of feed. The capacity of plate evaporator can easily be altered by changing the number of plates. It is also easy to clean. It is unsuitable for very viscous

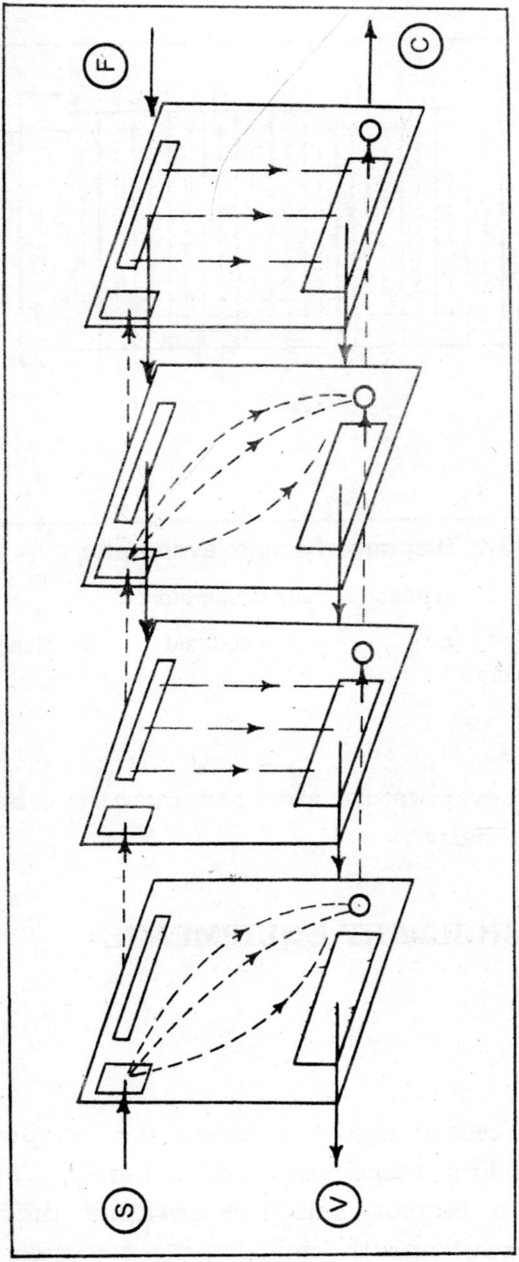

Fig. 3.10 Diagram of the flow of material in a plate evaporator.

C = Condensate, F = Feed, S = Steam for heating, V = Vapour and concentrate.

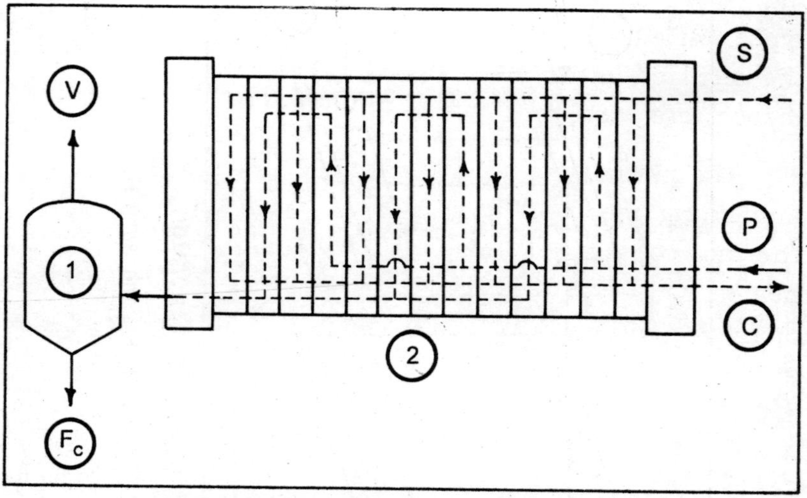

Fig. 3.11 Diagram of a plate evaporator

1. Separator, 2. Plate evaporator

V = Vapour, F = Feed, F_c = Concentrate, C = Condensate, S = Steam for heating

products. Plate evaporators are used more in other food industries than dairy.

ANCILLARY EQUIPMENTS

Pre-heaters

Milk is preheated before it enters the evaporator to ensure that boiling takes place immediately. The pre-heaters used to employ small temperature differences between the vapour and the milk, and are usually either multipass shell-and-tube or multipass concentric tube types. Milk flows through the tubes and the vapour inside the shell or outer tube. High velocities of milk 2 to 4.5 m/sec. are employed.

The pre-heaters are designated based upon the source of heating medium. A pre-heater using steam from the thermo compressor discharge is called on extraction heater; a heater drawing vapours from the inter-effect lines on multiple effect evaporators is called a bleeder, heater; and a vapour heater uses vapour from the last effect of an evaporating plant. Vapour heater is useful only for cold milk as the vapour temperature is only 45-50°C.

Condenser

Evaporators in the dairy industry run at temperature ranging from 40-70°C and have, therefore, to be operated under vacuum. To create a vacuum, vapour must be condense. The condenser serves to condense the vapour coming out from the last effect. Heat is transferred from vapours to the cooling medium. There are two principle types of condensers—spray type or direct contact and surface condenser (Fig. 3.12). The surface condensers are usually of shell and tube construction and costlier than direct contact types.

Vacuum Producing Equipment

Additional vacuum pump, such as stem jet air suction pumps, liquid seal pumps and others have only the purpose of removing non-condensable gases (e.g., air

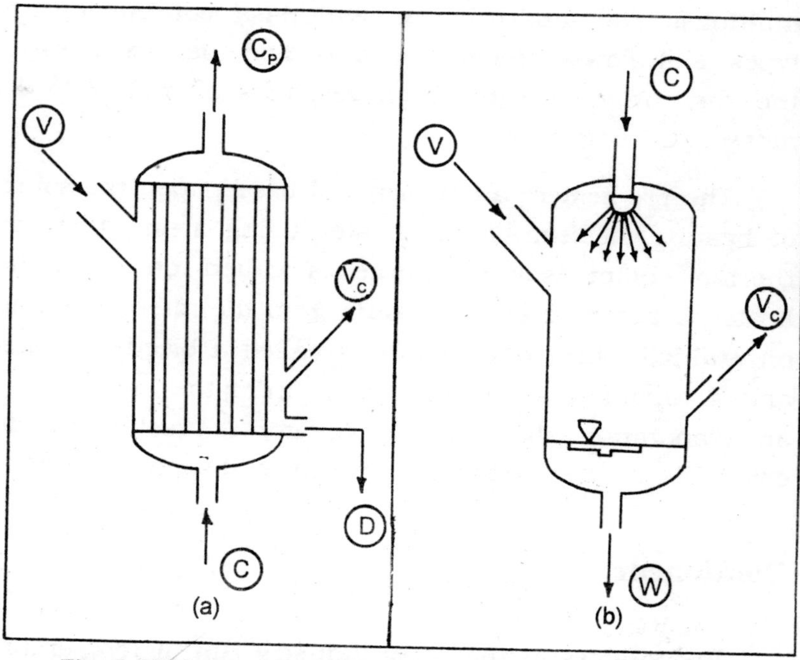

Fig. 3.12 Vapour condensers. (a) Surface condenser, (b) Spray type or direct contact condenser.

V = Vapour, V_C = Vacuum connection, C = Cooling water inlet, C_P = Cooling water discharged, D = Condensate, W = water condensed vapour mixture

which has leaked into the plant) and, thus, maintaining the vacuum.

Entrainment Separator

The concentrated milk and vapour mixture flow down the evaporator tubes into the entrainment separator where milk and entrained droplets of milk in vapour are separated from vapour. The most commonly employed types of separator is centrifugal or cyclone separator.

DISCUSSION QUESTIONS

1. Define evaporation. How is it different from drying ?
2. How will you select an evaporator ? Discuss.
3. Explain different types of evaporators used in dairy industry.
4. What do you understand by MVR ? What are the advantages of MVR ?
5. Describe plate type evaporators.

4

Milk Powder

Shelf-life is the catchword in today's dairy industry which is synonymous to milk powder. Here exists the supremacy of milk powder over its counterparts, i.e., other milk products, which are obviously having a low keeping quality due to higher rate of water activity. Milk powder contains less than 4% water, which decreases the volume of milk and resultant economic transportation of the product. Drying is, therefore, a method of preservation by which products are changed to a form which is light and capable of being stored at room temperature. The main purpose of powder manufacture is that it resembles as much as possible the original quality of milk after reconstituting.

Definition

Dried milk or milk powder is the product obtained by the removal of water from milk by heat or other suitable means to produce a solid containing 5% or less moisture. Whole milk, defatted milk or skim milk may be used for drying. The dried product obtained from whole milk is called whole milk powder (WMP) and that from skim milk is known as skim milk powder (SMP) or non-fat-dry milk (NFDM).

(a) Whole Milk powder—According to the PFA Rules, whole milk powder is the product obtained from cow or buffalo milk or a combination thereof from standardized milk by the removal of water.

(b) Skim milk powder—According to the PFA rules, Skim milk powder is the product obtained from skim milk of cow or buffalo or a combination thereof, by the removal of water.

Objectives of Preparing Milk Powder

- Handling excess milk supply in a dairy factory during the flush season.

- Optimal conservation of natural properties that are the characteristics of fresh milk.
- Reduction to minimum bulk to facilitate economic and easy transportation to all parts of the world.
- Extend shelf-life of the product.
- To provide a product to be used for other foods like ice-cream, infant food and bakery product.

Legal Standard

(a) Chemical Specification : The chemical composition and other specifications recommended for milk powder by ISI (Indian Standards Institution), PFA (Prevention of Food Adulteration Act) and ADMI (American Dairy Milk Institute) are as given in Table 4.1 and Table 4.2.

(b) Microbiological Standards : The microbiological limits set for different types of milk powders by Indian (ISI and PFA) and American (ADMI) agencies are given in Table 4.3 and Table 4.4.

Composition

The average percentage composition of whole milk and skim milk powder is given in Table 4.5. Salt Constituents, vitamins and Amino acids are given in Table 4.6.

Table 4.1

Chemical and Other Related Standards for Milk Powder as Prescribed by Different Agencies (ISI and PFA)

Characteristics	Requirement for	
	WMP	SMP
A. ISI		
Flavour and Odour	Good	Good
Moisture (% wt.) (maxi)	4.0	5.0
Total milk solids (% wt.)	96.0	95.3
Total ash (on dry basis) (%wt.) (maxi)	7.3	9.3
Fat (% wt.)	<26.0	>1.5
Solubility :		
(a) Solubility index (ml) (maxi)		
—Roller dried	15.0	15.0
—Spray dried	2.0	2.0
(b) Solubility (%) (mini)		
—Roller dried	85.0	85.0
—Spray dried	98.5	98.5
Titratable acidity (% lactic acid)	1.2	1.5
B. PFA		
Moisture (% wt.)	>5	>5
Fat (%wt.)	<26	>1.5
Additives (% wt.)	>0.3	>0.3
Titratable acidity (% lactic acid)	>1.2	>1.5

Table 4.2

Chemical and Other Related Standard for Milk Powder by American Dairy Milk Institute (ADMI)

Characteristics	Requirement for				
	Instant	Spray		Roller	
	US extra grade	US extra grade	US standard grade	US extra grade	US standard grade
Skim milk powder					
Moisture (%) >	4.25	4.0	5.0	4.0	5.0
Milk fat (%) >	1.25	1.25	1.5	1.25	1.50
Titratable acidity (%) >	0.15	0.15	0.17	0.15	0.17
Scorched particles (mg/g)	15.0	15.0	22.5	22.5	32.5
Solubility index (ml) >	1.0	1.2	2.0	15.0	15.0
Whole milk powder					
Moisture (%) >	2.25	2.25	3.0	3.0	4.0
Milk fat %) <	26.00	26.00	26.00	26.00	26.00
Solubility index (ml) >	0.5	0.5	1.0	15.0	15.0
Scorched Particles (mg/g)	15.0	15.0	22.5	22.5	32.5
Titratable acidity (%) >	0.15	0.15	0.17	0.15	0.17

Table 4.3

Microbiological Standards for Milk Powder as Prescribed by Different Agencies (ISI & PFA)

Test	Required (Counts/g) for	
	WMP	SMP
A ISI		
Bacterial Count (maxi)	50,000	50,000
Coliform count (maxi)	90	90
B. PFA		
Bacterial count (maxi)	50,000	50,000
Coliform count (maxi)	90	90

Table 4.4

Microbiological Standards for Milk Powders as Prescribed by ADMI

Test	Requirement (counts/g) for				
	Instant	Spray		Roller	
	US extra grade	US extra grade	US standard grade	US extra grade	US standard grade
Skim milk powder					
Bacterial count (maxi)	35,000	50,000	1,00,000	50,000	1,00,000
Coliform count (maxi)	90	—	—	—	—
Whole milk powder					
Bacterial count (maxi)	30,000	50,000	1,00,000	50,000	1,00,000
Coliform count (maxi)	90	—	—	—	—

Table 4.5

Composition of Whole Milk Powder and Skim Milk Powder (Main Component) in 100 g.

Constituent	Whole Milk Powder		Skim Milk Powder	
	Average	Range	Average	Range
Water	3.5 g	1.91-4.049	4.30 g	2.87-5.85 g
Protein (N × 6.38)	25.2 g	23.7-26.59	35.0 g	33.9-35.6 g
Fat	26.2 g	24.6-26.89	0.97 g	0.5-1.5 g
Carbohydrates	38.1 g	—	51.9 g	49.1-55.2 g
Ash	7.0 g	—	7.80 g	7.49-8.0 g

Table 4.6

Composition of Whole Milk Powder and Skim Milk Powder (Salt constituents, Vitamins & Amino Acid in 100 g).

Constituent	Whole milk powder		Skim milk powder	
	Average	Range	Average	Range
1	2	3	4	5
Salt Constituents				
Sodium	371 mg	302-410 mg	557mg	210-935 mg
Potassium	1.16 g	1.10 – 1.20 g.	1.58 g	1.00-2.05 mg.
Maganese	—	—	—	5-700 mg
Calcium	0.92 g	0.80-1.00 g	1.29 g	0.96-1.89 g
Magnesium	—	—	—	—
Iron	0.70 mg	0.60-1.34 mg.	0.8 mg	0.6-1 mg.
Cobalt	—	—	—	0.24 mg.
Copper	0.23 mg.	—	29 mg.	0-450 mg.
Zinc	4.5 mg	—	—	—
Phosphorus	714 mg.	700-725 mg.	1.02 g	0.90-1.89 g
Fluoride	—	—	—	—
Chloride	—	—	—	—
Iodine	—	—	—	—
Vitamins :				
Vitamin A	0.18 mg.	—	12 mg	—
Carotene	0.23 mg.	0.22-0.24 mg.	0	0
Vitamin D	3.8 mg	—	0.025 mg	—
Vitamin E	—	—	—	—

contd...

1	2	3	4	5
Vitamin K	—	—	—	—
Vitamin B_1	0.27 mg.	0.19-0.30 mg	0.34 mg.	0.23-0.45 mg.
Vitamin B_2	1.26 mg.	0.70-1.46 mg.	2.18 mg.	1.80-2.63 mg.
Nicotinic acid	0.70 mg.	0.60-0.76 mg.	—	1.1-1.2 mg.
Pantothenic acid	2.70 mg.	—	3.45 mg.	3.39-3.50 mg.
Vitamin B_6	0.20 mg.	0.19-0.21 mg.	0.28 mg	—
Biotin	10 mg	—	14 mg.	—
Folic acid	40 mg	—	3.4 mg	2.5-3.9 mg
Vitamin B_{12}	3.5 mg	3.2-3.7 mg	2.2 mg	—
Vitamin C	2.20 mg	0-6.0 mg	2 mg	—
Amino acids				
Isoleucine	1.61 g	1.09-2.09 g	2.24 g	1.51-2.91g
Leucine	2.47 g	2.03-3.08 g	3.43 g	2.82-4.27 g
Valine	1.73 g	1.11-2.13 g	2.40 g	1.54-2.96 g
Methionine	0.62 g	0.40-0.97 g	0.86 g	0.56-1.34 g
Cystine	0.23 g	0.097-0.39 g	0.31 g	0.13-0.55 g
Phenylalanine	1.22 g	0.93-1.67 g	1.70 g	1.30-2.31 g
Tyrosine	1.28 g	1.01-1.58 g	1.78 g	1.41-2.19 g
Threonine	1.16 g	0.89-1.52 g	1.61 g	1.23-2.11 g
Tryptophan	0.35 g	0.24-0.46 g	0.49 g	0.34-0.64 g
Lysine	1.96 g	1.28-2.41 g	2.72 g	1.78-3.34 g
Histidine	0.66 g	0.31-1.22 g	0.92 g	0.43-1.70 g
Agrinine	0.92 g	0.68-1.24 g	1.28 g	0.94-1.72 g

Desirable Attributes

(a) Skim milk powder : Table 4.7. shows desirable attributes of roller dried and spray dried skim milk powder.

Table 4.7

Desirable Attributes of Skim Milk Powder

Attributes	Roller dried	Spray dried
Flavour	Slightly cooked or heated flavour	Fresh skim milk
Body & Texture	Coarse & Less homogeneous if not finely pulverized	Fine smooth and homogeneous
Colour	Cow milk — Yellowish white Buffalow milk — Chalky white without burnt particle.	

(b) Whole Milk Powder (Spray dried)

Flavour : Clean, rich, sweet and very pleasant in flavour.

Body & Texture : Fine smooth and homogeneous.

Colour : Uniform in colour, light yellow for cow and creamy white for buffalo without burnt particle.

Food and Nutritive value

Modern drying methods preserve nutritive value of milk. While there appears to be only a slight destruction of lysine in spray drying, the severe heat treatment of roller drying destroys more lysine. Dry whole milk is a good source of vitamin A, calcium and phosphorus. It is also a valuable source of riboflavin.

Role of Milk Constituents in Manufacture and Storage of Milk Powder

(i) Milk Fat

Milk fat may be present in dried milks or milk products either in the original globule, i.e., in emulsion form or as free fat, i.e., in de-emulsion form.

- Free fat makes the dry product greasy and leaves an oily film on the reconstituted milk.
- The unsaturated fatty acids of milk fat, e.g., oleic, play a profound role in oxidative deterioration. Free fat promotes oxidation.
- The phospholipids of milk appear to be responsible for the oxidized flavour.
- The rancid flavour of milk fat is primarily due to the liberation of butyric acid as a result of hydrolysis of glycerides through lipase action.

(ii) Milk Proteins

- Milk proteins are readily subject to changes during heat processing and, to some extent, during storage; the changes adversely affect their solubility.

- The heat treatment destabilizes the proteins by inducing first stage denaturation and then irreversible denaturation (coagulation) as it becomes more severe.
- Some destabilization also occurs slowly during prolonged storage.
- β-lactoglobulin, appear to be the source of sulfydryl compound associated with a cooked and caramelized flavour.

(iii) Milk Sugar

In freshly dried milk, the lactose is still present in the amorphous or glass (non crystalline) state. In this form, it is very hygroscopic and readily absorbs moisture upon exposure to the atmosphere, which induces crystallization. This absorbed humidity then causes the powder to become sticky and then to cake.

- Browning discolouration of milk powder may be caused during manufacture, chiefly due to excessive temperature and long exposure to heat.
- Browning discolouration may also occur during storage; especially under conditions of high moisture, high storage temperature.

(iv) Milk Salt

- Milk salt plays an important part in influencing the physical state and physical stability of milk proteins, particularly caseinates.

- Minor changes in salt balance and pH easily upset the equilibrium and tend to destabilize and precipitate the casein particles. The resulting milk powder will have low solubility.

Shelf-life of Milk Powder

Skim milk powder has a maximum shelf-life of about three years, while whole milk powder has a maximum shelf-life of six months. This is because the fat in the powder oxidises during storage, with consequent deterioration in flavour.

However, the shelf-life can be extended by using appropriate packaging technology, i.e., packing the powder under an inert gas such as nitrogen.

Uses of Milk Powder

Milk powder can be used for various applications, such as

Recombination of milk : The powder is used in human nutrition, either as recombination of milk or milk product.

In the Bakery Industry : In the bakery industry, to increase the volume of bread and improve its water-binding capacity, the bread will then remain fresh for a longer period of time. The addition of about 6% (2-12%) of skim milk powder to the flour is recommended. In confectioneries and cakes, skim milk powder is added in amounts between 5 and 25%; in biscuits and related products, the quantity of skim milk powder added varies from 2 to 25% of the flour.

Chocolate Industry : In producing milk chocolate in chocolate industry, 5-18% skim milk powder is added. Roller-dried whole milk is normally used in milk chocolate manufacture (Hansen & Hansen, 1990).

Dairy desserts, frozen deserts & Ice-cream : Skim milk powder is also used in the production of dairy desserts, frozen desserts and ice-cream.

- Substitute for eggs in bread and pastries.
- In toppings, coatings and puddings, the amount of skim milk powder may be as high as 50%. Producing sausages and various types of ready-cooked meals in the food industry and catering trade.
- Use in baby foods.
- Animal feed, calf growth accelerator.

Technology of Whole Milk Powder Production

Production of whole milk powder consists of the following operations : reception, clarification, cooling and storage, standardization, fore-warming, evaporation, homogenization, drying and packaging (Fig. 4.1).

Raw Material : Milk used for the production of powder must be of high chemical, organoleptic and bacteriological quality. Very strict demands are made of the quality of the raw material for the production of milk powder.

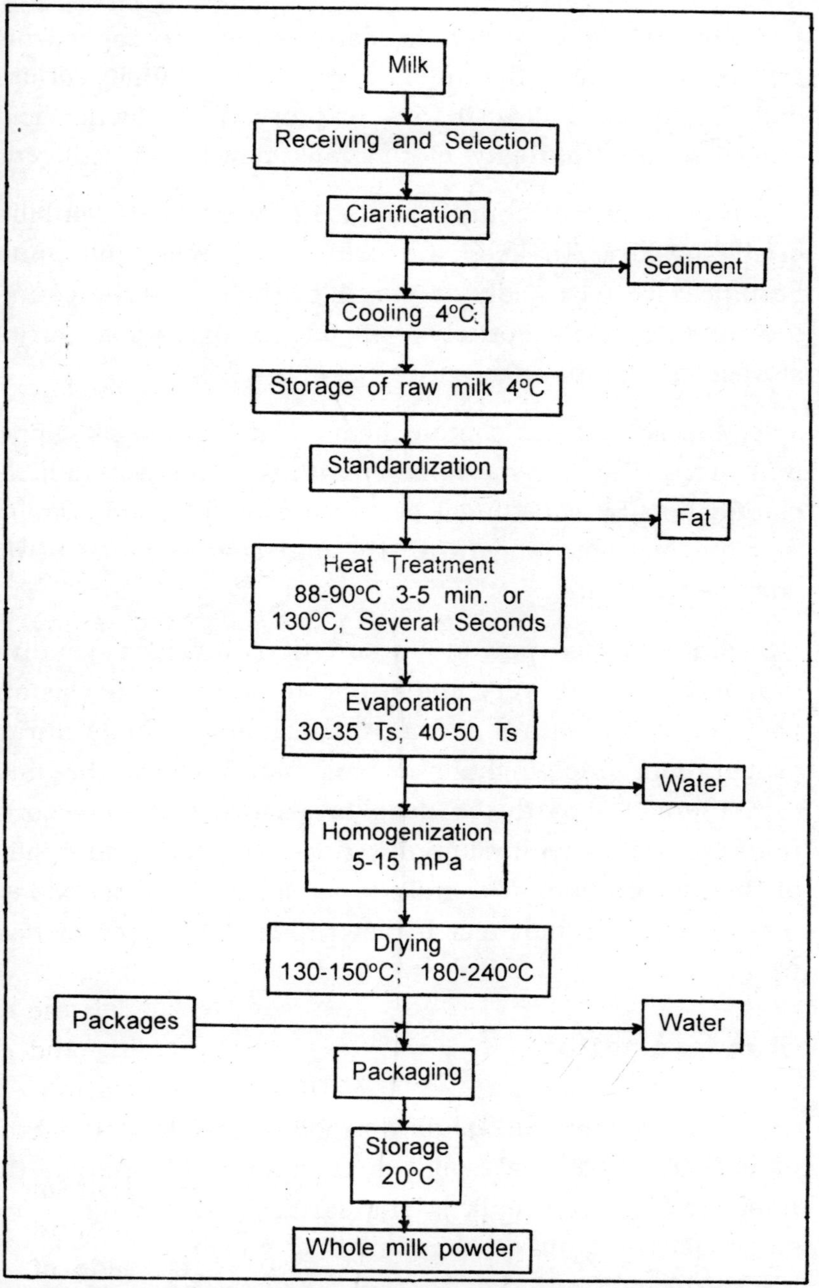

Fig. 4.1 : Flow chart of whole milk powder production

Chemical and bacteriological analysis are carried out at the time when the milk is received. Acidity of the milk must be below 0.15% (expressed as lactic acid) otherwise the solubility of the milk powder is reduced.

High bacterial count in milk decreases the solubility of the powder. To keep the count low, fresh milk must be processed as rapidly as possible. High bacterial counts increase the susceptibility of fat to oxidation during storage of the powder.

Milk should not contain heavy metals such as copper and iron, the conventional treatment of raw milk is clarification by centrifugal separators or filtration, cooling in plate heat-exchangers to 4°*C and storage tanks at the same temperature.*

Since spray powder production involves vacuum evaporation, it is very important to keep heat-resistant bacteria under control so that they cannot multiply during evaporation. Bactofugation or microfiltration of the milk can, therefore, partly be used to remove bacteria spores from the milk, thereby improving the bacteriological quality of the end product. The milk is subjected to a peroxidase test or whey protein test to determine the degree of heat treatment.

Standardization

The objective of standardization is to adjust the ratio of milk fat and total solids to the level required in the final product. Raw milk is standardized by adding to it a calculated amount of skim milk or cream.

Forewarming

The main objectives of forewarming are :

- To destroy all pathogenic bacteria and most of the saprophytic micro-organisms.
- To inactivate enzymes, especially lipase, which could provoke lipolysis during storage.
- To activate SH groups of β-Lactoglobulin, thus increasing resistance to the powder to oxidative changes during storage.

For milk powder solubility, which mostly depends on the state of the milk protein system, the High temperature short-time (HTST) is more convenient when compared to prolonged thermal treatment at low temperatures and has the same microbiological effect. The temperature most frequently used is high pasteurization in the range of 88-95°C for 15-30 seconds. Heat-treatment is commonly performed using the indirect method, in tubular or plate heat exchangers.

Good sanitation practices at each step are essential to reduce the microflora of milk powder.

Evaporation

Evaporation is a compulsory step in powder processing for several reasons.

- Evaporation process minimizes energy cost.
- Milk powder produced from evaporated milk has longer shelf-life and larger powder particles with a smaller amount of occluded air.

- The increased viscosity caused by higher concentration of total solids results in a powder having particle size.

Evaporator use : For concentrating milk prior to drying, a continuous multiple-effect evaporator, usually of the tubular type is used. Contemporary evaporators, in order to minimize energy costs, have six or seven effects or devices with TVR or one or two effects and MVR.

Concentration Rate : The maximum possible concentrations during evaporation is caused by drying technique.

(a) Roller drying : The concentration during evaporation is increased to 33-35% total solids, for roller drying technique. A higher concentration rate during roller drying would form a thick layer on the rollers, followed by inhibited drying and intensive irreversible changes to protein, lactose and fat.

(b) Spray drying : The concentration during evaporation is increased upto 40-50% for spray drying. However, further concentration for spray drying would increase viscosity and cause difficulties during atomization of the milk.

Homogenization

This step is usually incorporated in the roller drying process. The following objectives of homogenization are :

- To accomplish uniform distribution of fat in the whole milk powder.
- To decrease free fat content. Fat globules without protective membranes reduce milk powder

solubility and increase susceptibility to oxidative rancidity. During homogenization, free fat is transformed into fat globules due to protein absorption and fat globule membranes are formed.

- To facilitate reconstitution of dried milks to fluid milk without homogenization, the fat may churn during agitation while combining it with water.
- To improve the shelf-life of the powder.

Process : Homogenization is conducted after evaporation or in partly concentrated milk, the concentration rate being not more than 3 : 1. At higher concentrations, homogenization destabilizes milk protein, this decreasing powder solubility. Homogenization pressure is 5-15 MPa.

In plants, where milk powder is produced by mixing cream and skim milk, only cream should be homogenized. However, in practice, the final concentrate is usually homogenized.

To avoid contamination during homogenization the homogenizer has to be properly cleaned and sanitized.

Drying

Industrial scale, the milk is commonly dried by application of roller dry or spray drying method.

Roller drying : In roller drying, the milk concentrate is distributed as a film on rotating steam-heated drums or rollers. As the water evaporates, a thin layer of dried milk is formed on drum that is continuously removed by

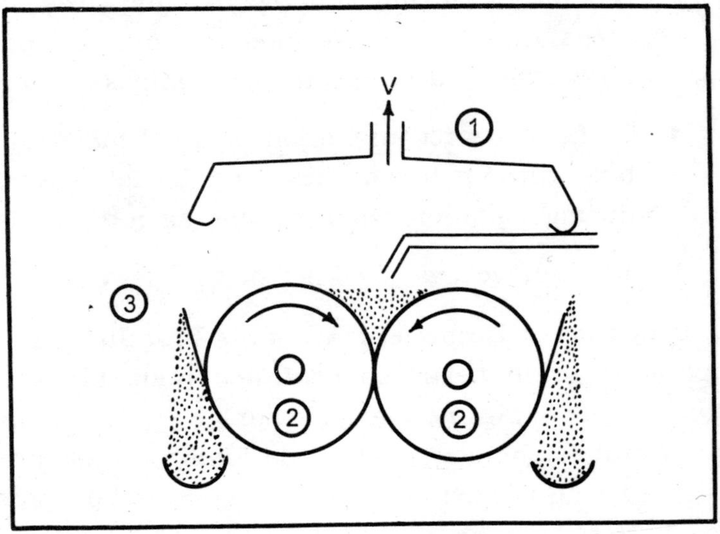

Fig. 4.2 Diagram of roller dryer

1. Vapour hood, 2. Drum, 3. Knife, V = Vapour, F = Feed

a scraper (Fig. 4.2). The product is in the form of flakes, scales, which are then transferred to a grinder. Grinder is used to create desired grain size of product. After that, hard and burned particles are shifted off. Depending on the desired capacity, the roller dryer is 1-6 m. long and has a roller diameter of 0.6-3 m. Its performance depends on film thickness, roller surface temperature, roller speed and the DM content of supplied milk (Details given in Chapter 5).

Vacuum roller drying : Drying of the milk at a temperature below 100°C at 91-98 kPa means that the effects of oxygen and high temperature are markedly reduced. The product has better properties than the milk powder obtained from roller drying under atmospheric pressure.

Disadvantages : Direct contact of the preconcentrated milk with the hot drum surface results in some irreversible changes in the milk. Caramelization of lactose, Maillard's browning reactions and denaturation of proteins are the most important changes affecting the quality of roller-dried milk powder. Denaturation of proteins decreases the solubility of the powder.

Advantages : Low cost investments

- Relatively small space necessary for the operation
- Economical processing.
- Easy operation and maintenance.
- Intensive heat treatment increases the water binding properties of the powder. This characteristic is useful in the production of ready-cooked meals, sausages and bread, cakes and pastries.
- Roller powders are used in chocolate industry in particular due to their special properties.

Factors Effecting the Quality of Roller-dried Milk Powder

- The temperature and duration drying
- The temperature and concentration of milk feed.
- The thickness of the milk layer over the rollers.
- Uniformity of the milkfeed.

Spray Drying

The spray drying process, in recent years, has become the acceptable method of drying milk. The roller drying process is limited to the products for special use. In India, most of the milk drying plants are spray dryers.

Principle : The concentrated milk is turned into powder in a spray tower. (Fig. 4.3). Drying process is also a multiple-stage process.

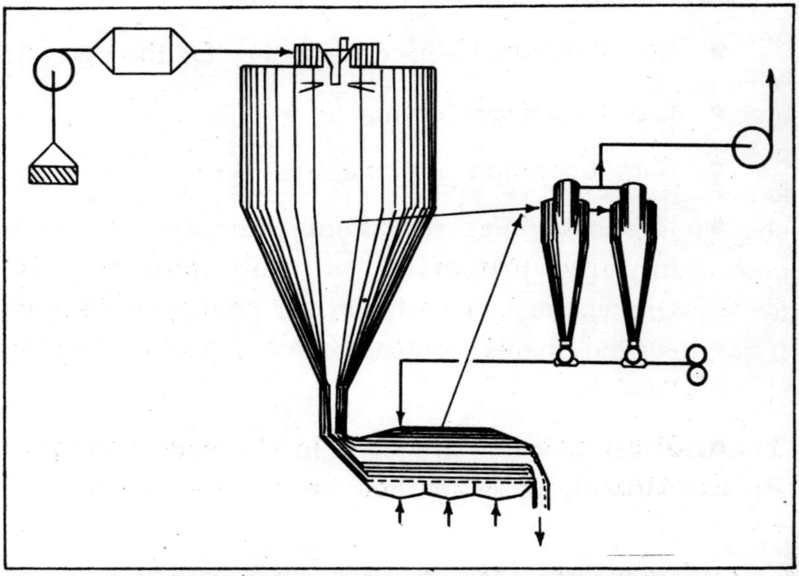

Fig. 4.3 Spray dryer : drying chamber with conical bottom and fluid-bed

(***Courtesy*** : APV ANHYDRO A/S, Denmark.)

- Atomization of concentrate into very fine droplets in a hot air stream.
- Water evaporation
- Separation of the powder from the drying air.

The concentrated milk is dried with inlet air at 143 to 232°C. and exit air at 74 to 93°C, depending on product characteristics. To reduce heat damage during drying and yet obtain the desired moisture, a low exhaust air temperature is preferred.

The time necessary for a milk droplet to be dried in a spray drier depends mainly on droplet diameter and air temperature. Dry product falls to the chamber bottom and is continuously discharged from there.

Cooling : The whole milk powder should be removed promptly from hot air stream to maintain better flavour and body characteristics and also long shelf-life. The higher the temperature and the longer the time the product is above the melting point of fat, the greater the amount of free fat obtained, thus adversely affect the shelf-life. Hence, it is essential to cool the powder immediately to a temperature below the melting point of fat. Powder is cooled by a cold air stream and separated from the drying medium.

Technology of Instant Whole Milk Powder Production

Effort to improve the wetting properties of spray dried milk have led to the development of special processes for producing "instantly soluble" powder. Instantization refers to the process by which dried milk and milk products are made instant-soluble. The patent for this procedure was granted to Peebles in 1955.

Principles

Agglomeration is the most important methods of obtaining instant properties. The method depends on the principles of providing a structure of porous aggregates which disperse readily and quickly without vigorous agitation, to produce a stable reconstituted milk. This is achieved by slight moistening of the powder particles so that they develop surface stickness and adhere particles together as large porous mass of irregular shape after which the aggregates are partially re-dried to preserve their structure. (Fig. 4.4)

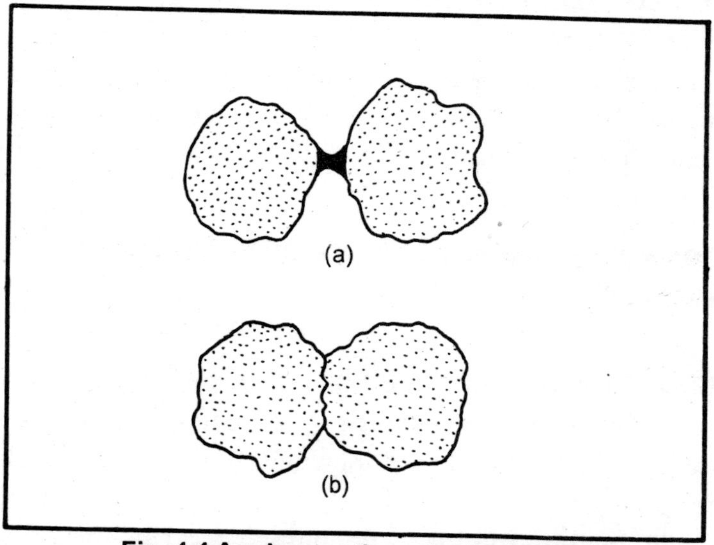

Fig. 4.4 Agglomeration of milk powder

Instantizing Affects the Ability of Dried Powder Particles

- To absorb water on its surface (wettability)
- To penetrate through a surface layer of water. (penetrability)
- To sink through water, after being moisturized (Sinkability).
- Disperse without formation of agglomerates (dispersibility)
- Dissolve quickly (rate of dissolving).

Instantization of whole milk powder : The technological process of instantization is more complicated if the product containing fat (e.g., whole milk powder). The surface of the particles in whole milk powder will normally be covered by a fat layer. It is the free fat on the surface of the particles which makes them water repellent (hydrophobic) and surface active agents have to be added to improve the moistening. Instantization of whole milk powder, as developed by Pisecky and Westergard, quantitates the controlled ratio of free fat to lecithin. If the powder particles get a coating containing lecithin, good moistening of the powder is ensured provided that the lecithen concentration and the quantity of the coating materials are right in proportion to the specific surface and free fat content of particles. The lecithination process is carried out either immediately after the agglomeration or as a single process on previously agglomerated powder.

Process

(i) The powder is two-stage dried and the fines are recirculated to the wet zone. The fluid bed is divided in three sections (Fig. 4.5) The first section is for drying of the powder. In the second section, the powder is lecithinated by adding lecithin mixed with butter oil. The amount of added lecithin is 0.3% of the powder. In the last section of the fluid-bed, the powder is stabilized to secure a good distribution of the lecithin on the surface of the particles. During lecithination and stabilization the powder temperature is very important factor. It is necessary to have butter fat on particle surfaces and the lecithin on a liquid state. Therefore, normally, the powder at temperature 45°C-50°C is used during lecithinating process.

(ii) Fig. 4.6 shows a plant for lecithination of previously agglomerated whole milk powder. The powder is taken from a silo. In the first section of the fluid bed the powder is heated to about 50°C for melting the fat on the surface of the particles. The lecithination and the stabilization of the lecithin are carried out in the second and third section, respectively. The procedure will be the same as mentioned earlier.

Parameters in the production : A critical parameter in the production of instant whole milk powder is the state of the fat in the powder particles. The fat should be protected in the particles to keep the content of free fat low, normally not more than 2% of the total fat content. Higher content of free fat will cause a poorer wettability of lecithinated powder.

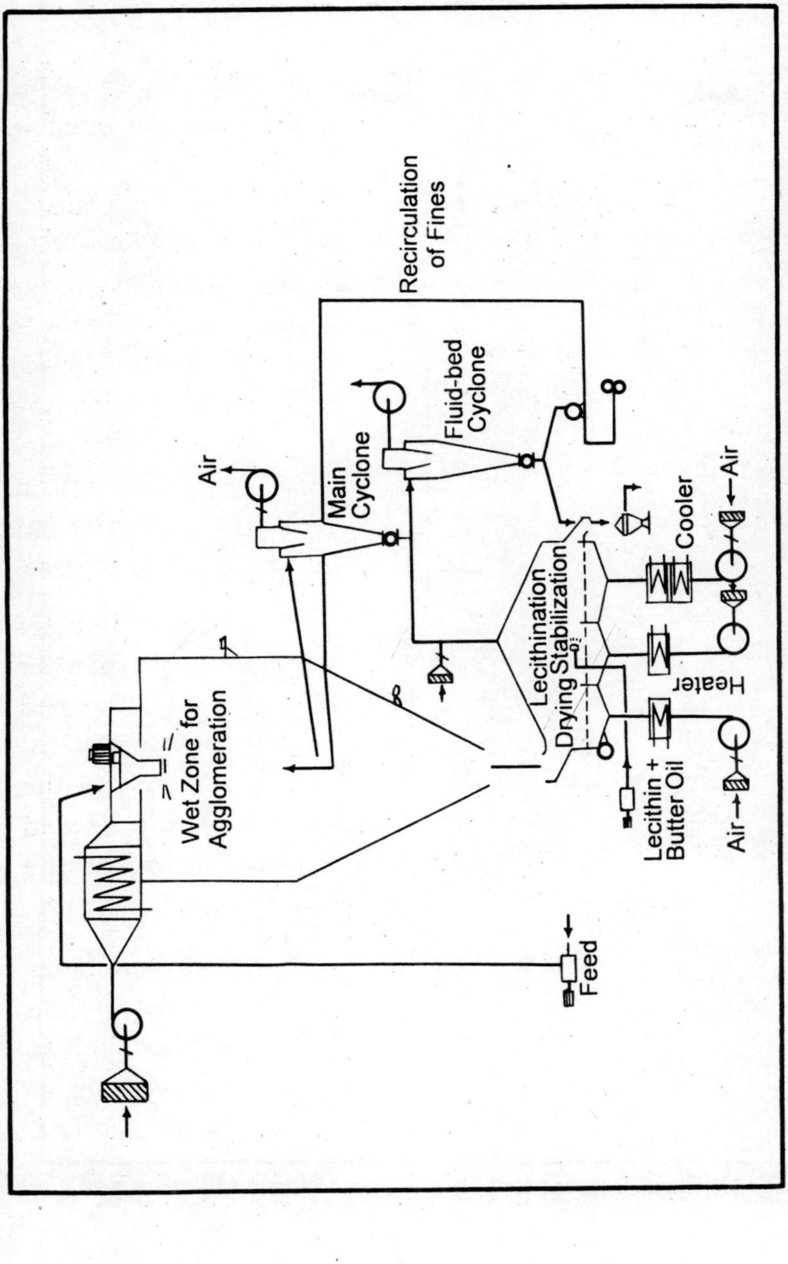

Fig. 4.5 Spray-Plant for straight-through agglomeration and lecithination of whole milk powder

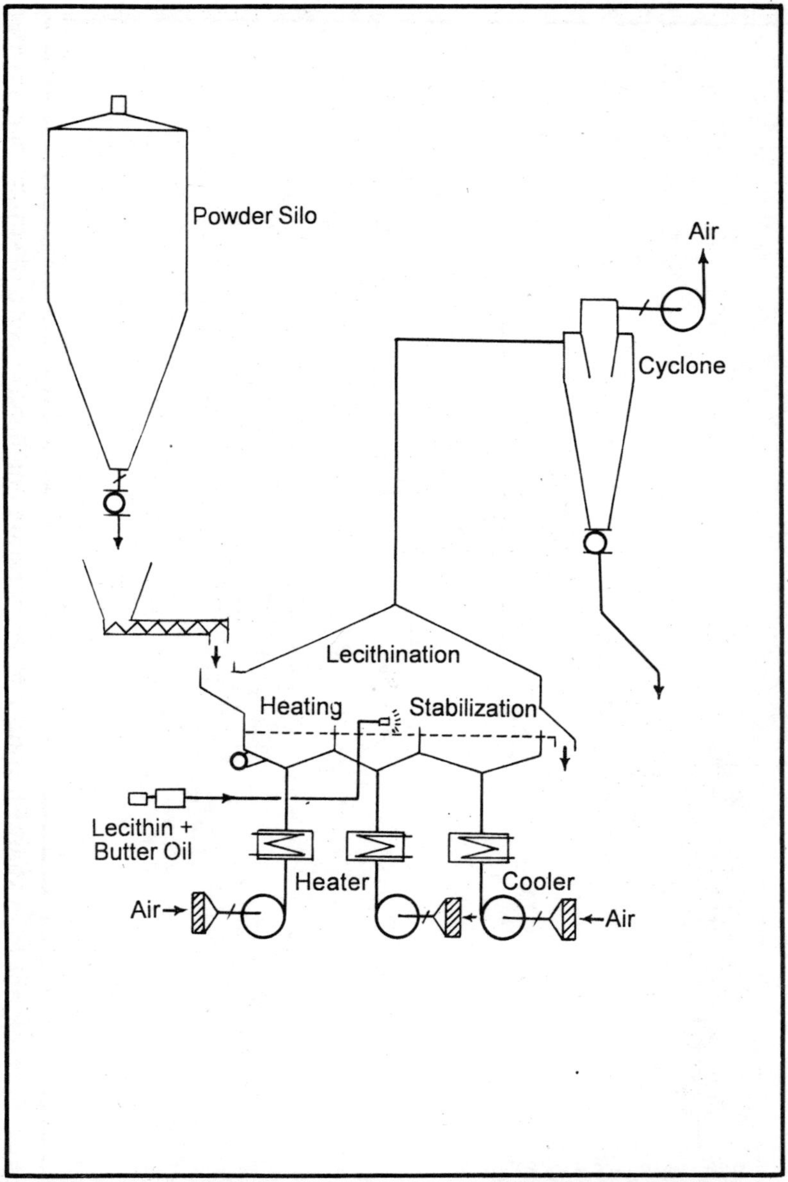

Fig. 4.6 Fluid Bed for Lecithination of Whole Milk Powder

The homogenization of whole milk concentrate and content of vacuoles and capillaries in the powder particles should be as low as possible. This is influenced by the total solid in the concentrates and the drying temperature. The homogenization should be as effective as possible. The pressure should be high, but not as high as to increase the viscosity of concentrate to a large extent.

Packaging : The lecithinated powder should be packed in a suitable material to protect the agglomerates from breaking down during transport and handling. Tins are best packing warn and cooling should be carried out in the tins placed in the cold room. The powder should be cooled to temperature below the melting point of the butter fat; usually to below 30°C and during storage and transportation the powder should never be heated to temperature higher than the melting point.

Technology for Skim Milk Powder Production

Production of skim milk powder differs slightly from whole milk powder production. The basic operation are receiving and selecting milk, clarification, cooling and storing the raw milk, then separation, pasteurization, evaporation, drying, packaging and storing the powder (Fig. 4.7). The first part of the process, before reaching fat separation, corresponds to the operations used for whole milk powder production.

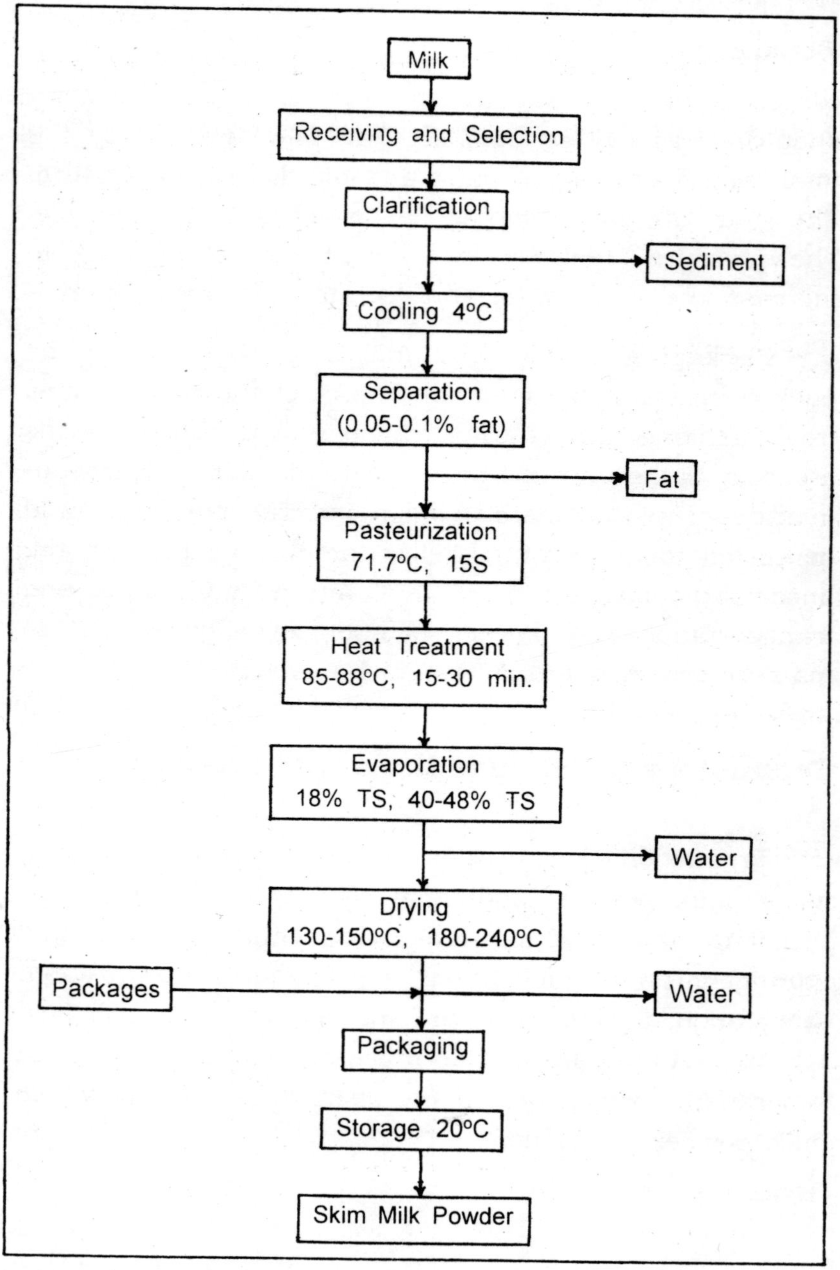

Fig. 4.7 Flow Chart of skim milk powder production

Separation

To obtain skim milk powder from whole milk, it is necessary to reduce fat in raw milk to 0.05-0.10%. This procedure is done by separators. Efficiency of separation is increased if the milk has been preheated to 27-32°C.

Pasteurization

Pasteurization is most intensive heat treatment in Skim milk powder production by the low-temperature method. The main goal of pasteurization are, destroy all pathogenic and most saprophytic micro-organisms and inactivate enzymes with minimal detrimental heat-induced changes such as serum protein denaturation. HTST pasteurization is accomplished at 71.7°C for 15 seconds in tube/plate heat exchangers. After pasteurization, skim milk is immediately pumped for evaporation.

Heat Treatment

Heat treatment is carried out only when skim milk powder is produced by a high heat, as opposed to a low-heat method. Heat treatment varies from 85 to 88°C for 15 to 30 minutes. The treatment is usually applied to the milk in the vessel called 'Hot well'.

Evaporation

Skim milk evaporation requires the use of plate or tubular type multi-effect evaporators as does whole milk powder production. If evaporated milk is spray dried, the total solid concentration after evaporation may be as much as 40-48%. it is only upto 18% when milk is roller dried. The concentration difference in the final product between skim and whole milk by both drying methods is caused by the relatively higher non-fat solids in skim milk, primarily proteins. Because of the higher concentration of proteins, the high temperature of the rollers creates irreversible protein changes. These changes are intensified by the presence of high milk salt concentrations.

Drying

Drying is carried out by spray dryer or roller dryer. Skim milk powder is susceptible to humidity and heat-induced changes. Heat induced changes can intensify the adverse effect on product quality. It must be cooled immediately after drying.

Removal of more water from skim milk during evaporation improves drying efficiency, thus increasing capability of the chamber.

Technology of Instant Skim Milk Powder Production

An effective agglomeration is, the most important way of achieving instant properties, and for majority of dried milk products it is the only mean available to achieve instant properties. The two principal methods are :

- Straight through method
- Rewet method

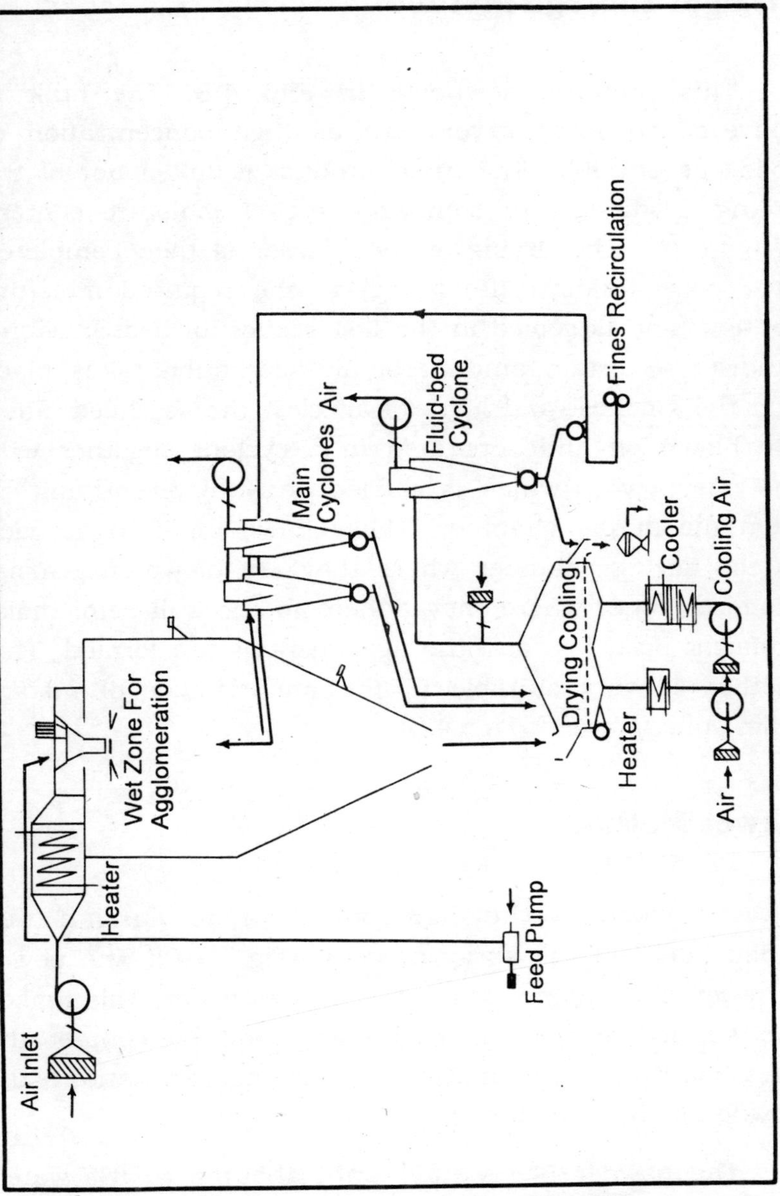

Fig. 4.8 Spray plant for straight through agglomeration

Straight through Method

This method is shown in Fig. 4.8. The milk is conveyed to spray dryer with as high concentration of solids as possible. The dried product is taken out of the drying chamber with somewhat higher moisture content of 6 to 7%. The drying of the powder is then completed in a vibro fluidizer (fluid bed) to the required moisture content, and is cooled in the last section of bed. In vibro-fluidizer a certain amount of air separation takes place and the smallest and lighter particles, the so called 'fines' are blown off and separated in a cyclone together with the fine powder which is carried away by the drying air from the drying chamber. These fines are brought back to the drying chamber where they are blown into moist atomizer zone, where they adhere to the still semi moist atomizer particles, and the agglomerates are formed. This method is very suitable in the manufacture of instant skim milk powder (Fig. 4.9).

Rewet Method

In process, the agglomeration is carried out independent of the spray process. (Fig. 4.10). The power is rewetted by using steam or atomized water. This makes the surface of the particles sticky and by contact the particles form agglomerates. After agglomeration, the powder is redried and cooled (Fig. 4.11).

The powder is rewetted until about 6 to 8% water. Higher water content should not be used, as there will be risk of reduction in the powder solubility. Lower water

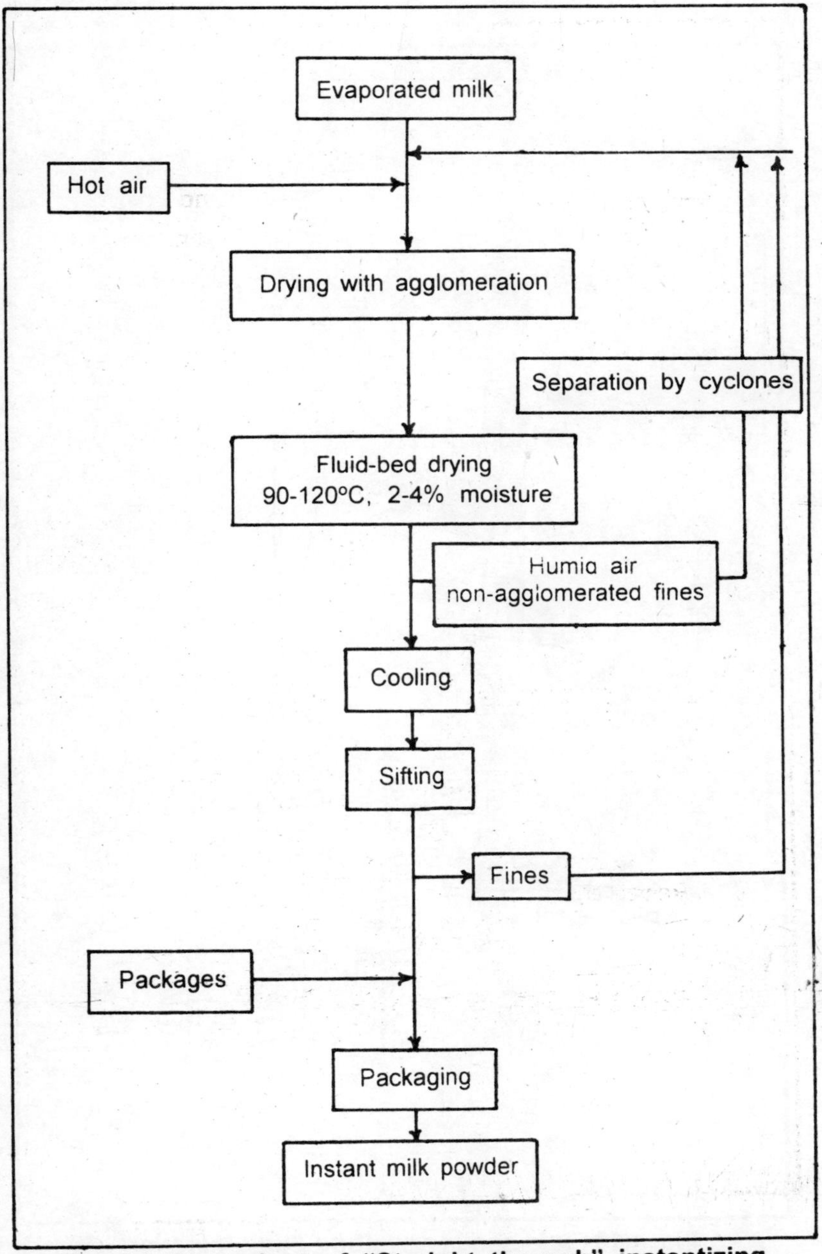

Fig. 4.9 Flow chart of "Straight through" instantizing procedure of instant milk power production.

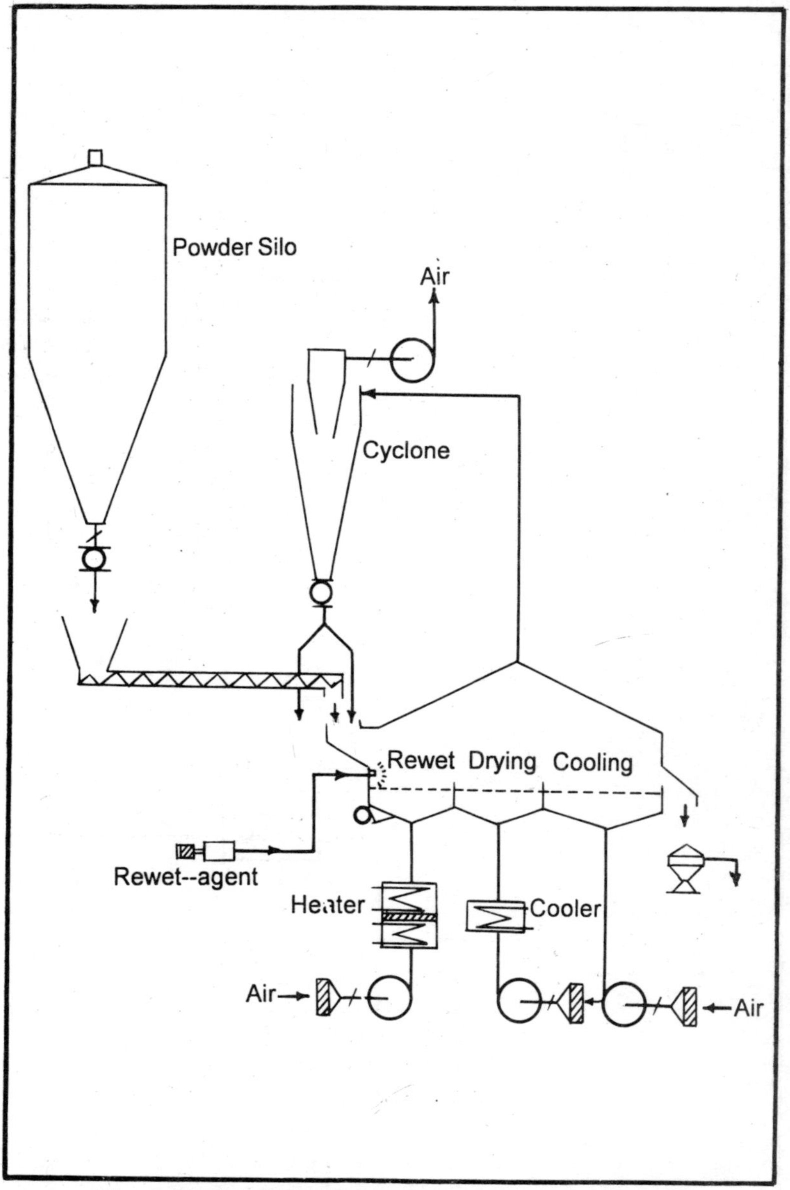

Fig. 4.10 Fluid bed in rewetting-agglomeration

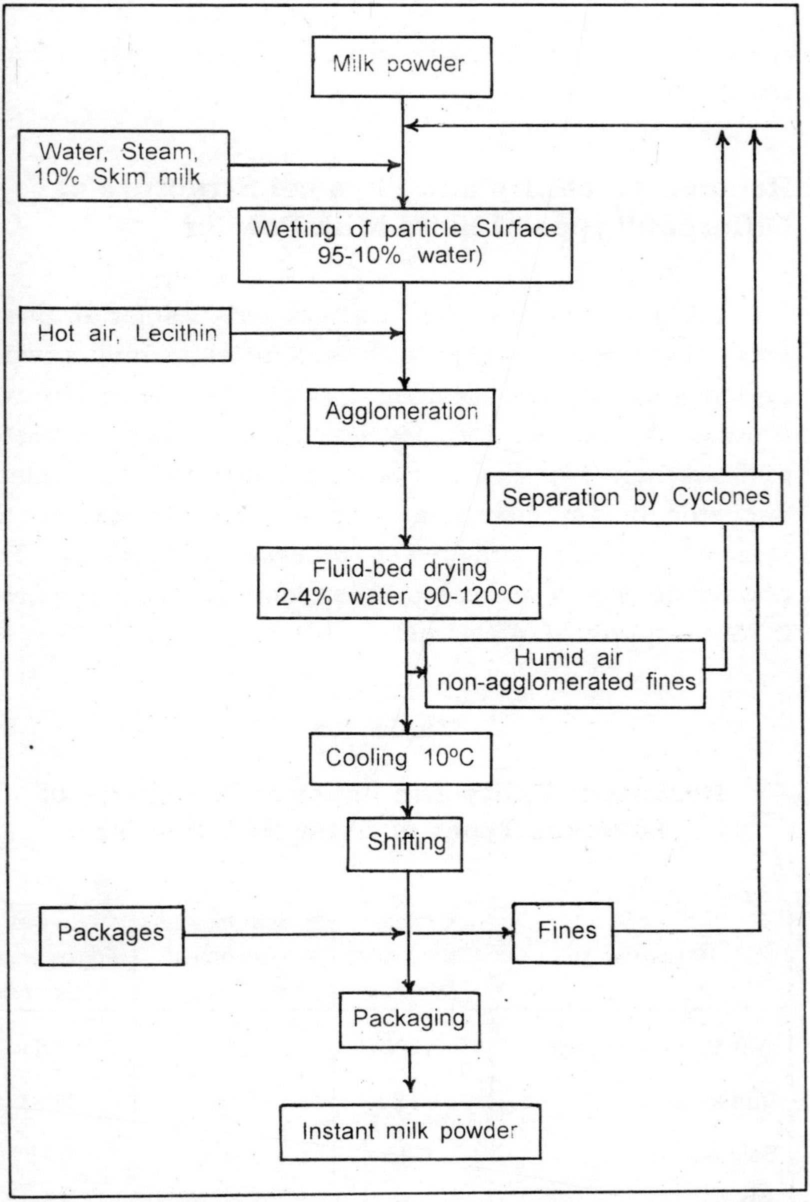

Fig. 4.11 Flow Chart of instantized milk powder production (rewet procedure).

content will cause a lower degree of agglomeration after rewetting.

Reconstitutability and Physical Structure of Different Types of Skim Milk Powder

Table 4.8 shows typical characteristics for ordinary spray dried skim milk powder and for skim milk powder agglomerated by the straight through process and re-wet process. The powder particles which have a mean diameter of less than 100 microns are agglomerated to a mean diameter of 250 microns and more than 400 microns for straight through and rewet process respectively. The powder density is decreased by agglomeration from approx. 0.70 to approx. 0.45 g/ml.

Table 4.8

Reconstituability and Physical Structures of Different Types of Skim Milk Powder

Properties	Ordinary Spray dried Skim milk	Straight through Agglomerated Skim milk	Re-wet Agglomerated Skim milk
Wettability (Seconds)	1000	20	10
Dispersibility (%)	60.80	90.96	92.98
Solubility Index	0.15	0.15	0.25
Average particles size (microns)	100	300	500
Density g/ml.	0.65-0.70	0.50-0.55	0.45-0.50

The instant properties are normally better in revert process than the straight through process because of more effective agglomeration. Unfortunately, the rewet process is more critical for the proteins in the milk to the effect that the solubility of rewet instant powder is very often poorer than that of straight through agglomerated instant skim milk powder. Further, the rewet process is more expensive from energy point of view and it is a more complex process compared with straight through process.

Fortification of Dry Milk Powder

Dry milk powder are relatively easy to fortify with vitamins particularly vitamin A. Dry blending of dry form of vitamin with milk powder is a suitable blend to ensure homogeneity. However, fortification can also be done in milk or milk concentrate prior to drying. The most widely practised fortification options are given in Fig. 4.12.

(a) Dry Blending option

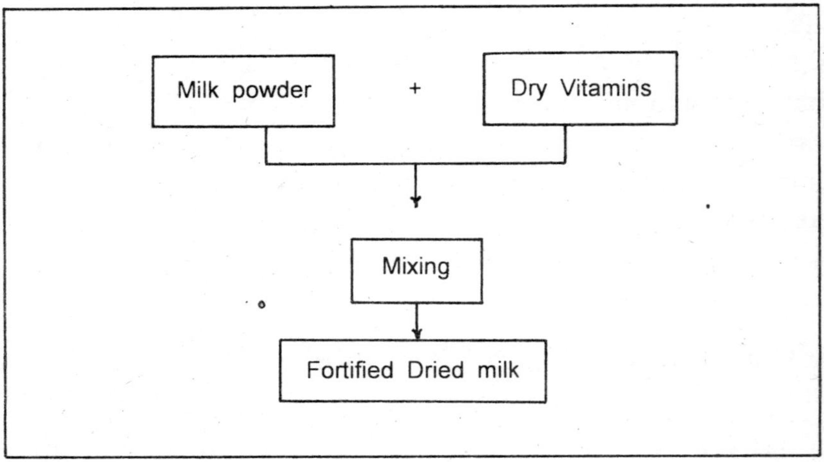

(b) Wet Blending Option

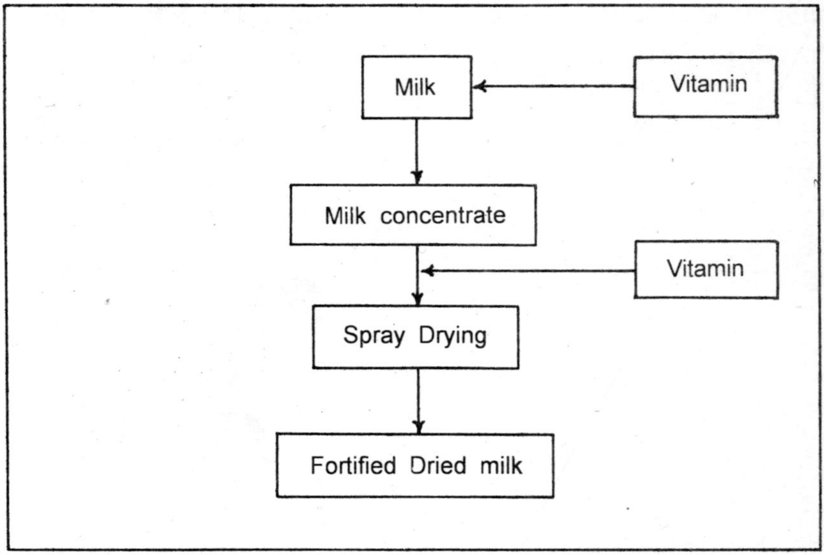

Fig. 4.12 : Fortification of Spray dried milk powder with vitamin A.

HEAT CLASSIFIED MILK POWDER

The concept of producing heat classified milk powder is comparatively new to India although it is widely in vogue in many countries. In fact, Milk fed, Punjab has already started heat classified powder manufactured by its different powder plants in Punjab, India. (Munjial, 1995). The production of milk powder to meet a certain heat classified was initiated in the U.S.A. Heat classification based on the amount of whey protein in the powder are issued by American Dry Milk Institute (ADMI, 1965). In New Zealand and Australia, practically all the powder is produced to meet a certain heat classification.

Heat classification basically refers to skim milk powder. The heat classification of Skim milk powder is done on the basis of level of denaturation of whey proteins. The desired degree of whey protein denaturation in the dried milk is obtained by variation of the pre-heating conditions to which the treatment of milk during processing is additive.

Composition of Milk protein and their specific properties : When skim milk that has not been heated, is acidified to pH 4.6 approximately 80% of the proteins get precipitated. These proteins are designated as casein and the mentioned pH value (pH 4.6) is called the isoelectric point. It is well known, that precipitation of the casein can also take place by the action of rennet or by saturation with sodium chloride. The proteins remaining infiltrate after the precipitation of the caseins are called whey or serum proteins and they constitute approximately 0.6% of the milk or 20% of total milk proteins. The composition of serum protein fraction based on the electrophoretic analysis is approximately 55% β lactoglobulin, 12% α-lactalbumin, 10% proteose peptone and 10 to 15% casein, the remainder being composed of globulins and enzymes. However, the two main groups of whey proteins are the lactalbumins and the lactoglobulins.

Denaturation of whey proteins : Whey proteins do not precipitate with acid unless they have been exposed to heat treatment to such an extent that their 'denaturation' indicates that the proteins are no longer in their natural configuration.

Characteristic of denatured whey protein :

- Change in the molecular configuration
- No damages regarding nutritional and physiological characteristics.
- The change is irreversible, i.e., when cooled, they do not again assume the original configuration.

Effect of heat on proteins : The effect of heat on proteins can be considered as a two stage process.

(i) The secondary and tertiary structure is altered, causing denaturation.

(ii) The protein aggregate, which may lead to coagulation, β-lactoglobulin is a well characterized protein which has been shown to undergo such process.

Whey Protein Nitrogen Index (WPNI)

Heat treatment denatures whey protein, the percentage denatured increases with the intensity of the heat treatment. The degree of denaturation is normally expressed by the whey protein nitrogen index (WPNI) measured in mg. of nitrogen/gm. of skim.

Method for Determination of WPNI

It has been published by ADMI (1965). The principle is as follows :

A sample of milk powder is reconstituted in distilled water. The casein and the denatured whey proteins are precipitated with sodium chloride and separated off by filtration. To the filtrate which contains the whey proteins, acid buffer is added and the turbidity produced by coagulation of the proteins is measured by means of spectrophotometer. The percent transmission reading obtained is connected to the amount of whey protein by means of a previously prepared standard curve (Fig. 4.13) which gives the relationship between optical density and level of whey proteins.

Categories of Heat Classified Skim Milk Powder

Depending on the intensity of the heat treatment, milk powder is classified into categories related to the temperature/tune combinations to which the skim milk has been exposed prior to and during evaporation and spray drying. Information about the various categories of spray dried skim milk powder is summarized in Table 4.9.

The heat treatment classification of non-fat dry milk as per ADMI is shown in Table 4.10.

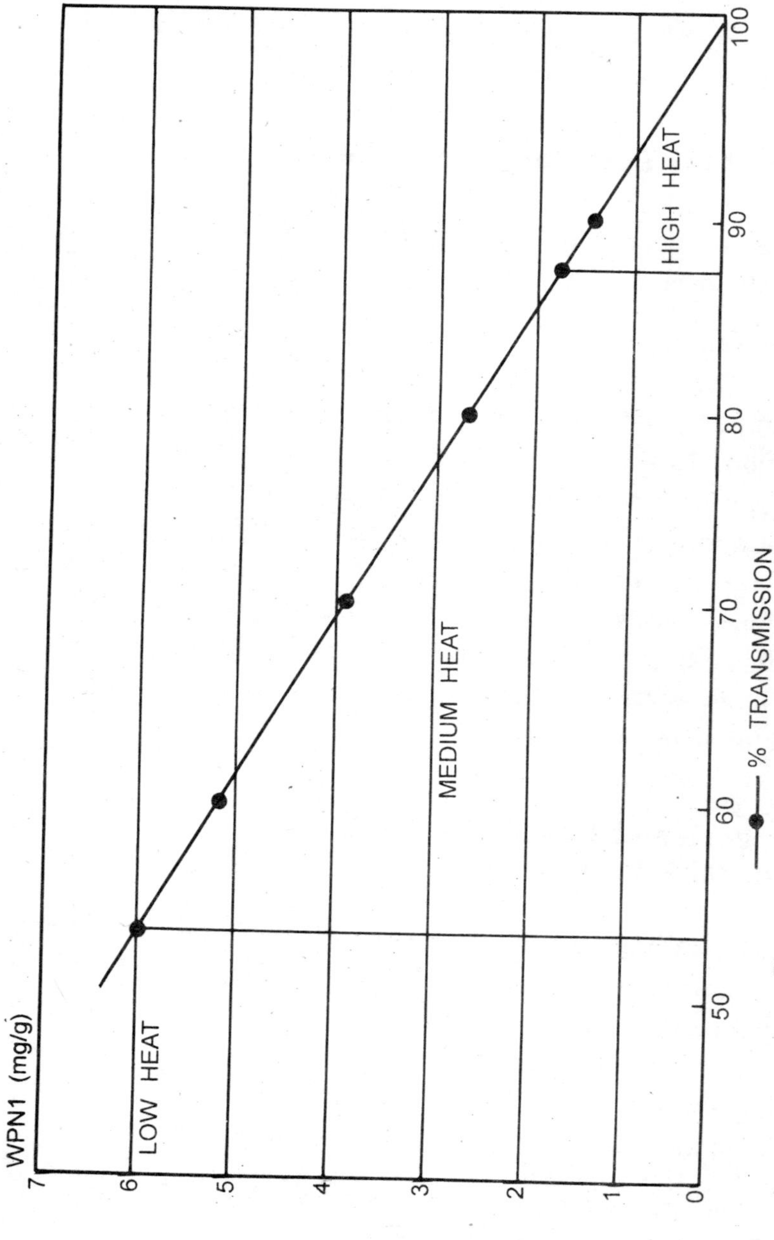

Fig. 4.13 Conversion curve for WPNI of Skim milk powder

Table 4.9

Categories of Spray-dried Skim Milk Powder

Category	Temp/time	WPNI mg/g
Extra low-heat	< 70°C	*
Low-heat (LH) Powder	70°C/15 S	> 6.0
Medium-heat (MH) powder	85°C/20 S	5-6.0
—Do—	90°C/30 S	4-5.0
—Do-	95°C/30 S	3-4.0
Medium high heat (HH)	124°C/30 S	1.5 – 2.0
High-heat (HH)	approx. 135°C/30 S	<1.4
High-heat high stable (HHHS) (from selected milk)	approx. 135°C/30 S	<1.4

* not measurable

Source : Sanderson (1967)

Table 4.10

Heat Classification of Non-fat Dry Milk as per ADMI

Class	WPN Index (mg g[1] – powder)
Low heat powder	≥ 6
Medium heat powder	> 1.5 < 6
High heat powder	≤ 1.5

Relationship between WPNI and Solubility Index

The solubility index is mainly a measure of the coagulated casein. It has been confirmed that the insoluble matter mainly consists of the coagulated casein with small amounts of fat and mineral complex by the analysis of the sediments obtained when solubility tests are made. The denatured whey proteins are only found in rare cases. There is no relationship between WPN index and solubility index and whey protein denaturation is not synonymous with insolubilisation.

Heat Treatment for Manufacture of Heat Classified Powder

In prescribing the heat treatment it is required to produce the powder of a certain WPN Index. It needs to be recognized that the heat treatment during the whole process is cumulative. The effect on whey protein denaturation of pasteurization, preheating prior to evaporation, during evaporation, concentrate heating and drying must be considered. It is easy to measure the effect of each heating stage by determining the WPN Index after each stage.

Pasteurization : The conventional pasteurization conditions required to ensure the bacteriological quality of the product have only a small effect, as a more severe heat treatment is required to cause any appreciable amount of whey protein denaturation. Only at times when the content of whey proteins in the raw milk is very low, and when low heat powder is to be manufactured. At this case gental pasteurization condition will be necessary. This type of pasteurization is acceptable when bacteriological standard of the raw skim milk is exceptionally high.

Pre-heating prior to evaporation : In this stage, whey protein denaturaion can take place. The WPN Index of the powder is in practice, governed by the pre-heating condition of milk before evaporation.

Effect of evaporator : The time of temperature conditions in evaporator vary according to the design of the plant. If whey protein denaturation is to be avoided during evaporation, the boiling temperature must not exceed 70°C and re-circulation where the product temperature is high, should be avoided.

It is useful to measure how the complete process, without pre-heating, alters the WPN Index from raw milk to powder. With this knowledge, experience will soon be obtained as to the pre-heating condition that results in powder of a certain heat classification. It is normally necessary to alter the pre-heating condition over the year.

Time temperament needed to achieve any desired WPN Index rate is given in Fig. 4.14. This should be treated as a guideline only. the actual treatment required depends on many factors, including the composition of milk as influenced by the breed, season, climate etc. and the equipment including type and construction of the heater.

Low heat non-fat dry milk manufacture requires that heating be carefully controlled to result in a minimum amount of heat induced changes and yet accomplish proper pasteurization. Content of both time and temperature assumes much significance. High temperature short time pasteurization as part of the pre-heating may be tubular or plate heater not less than 161°F (71.7°C) for at least 15 seconds.

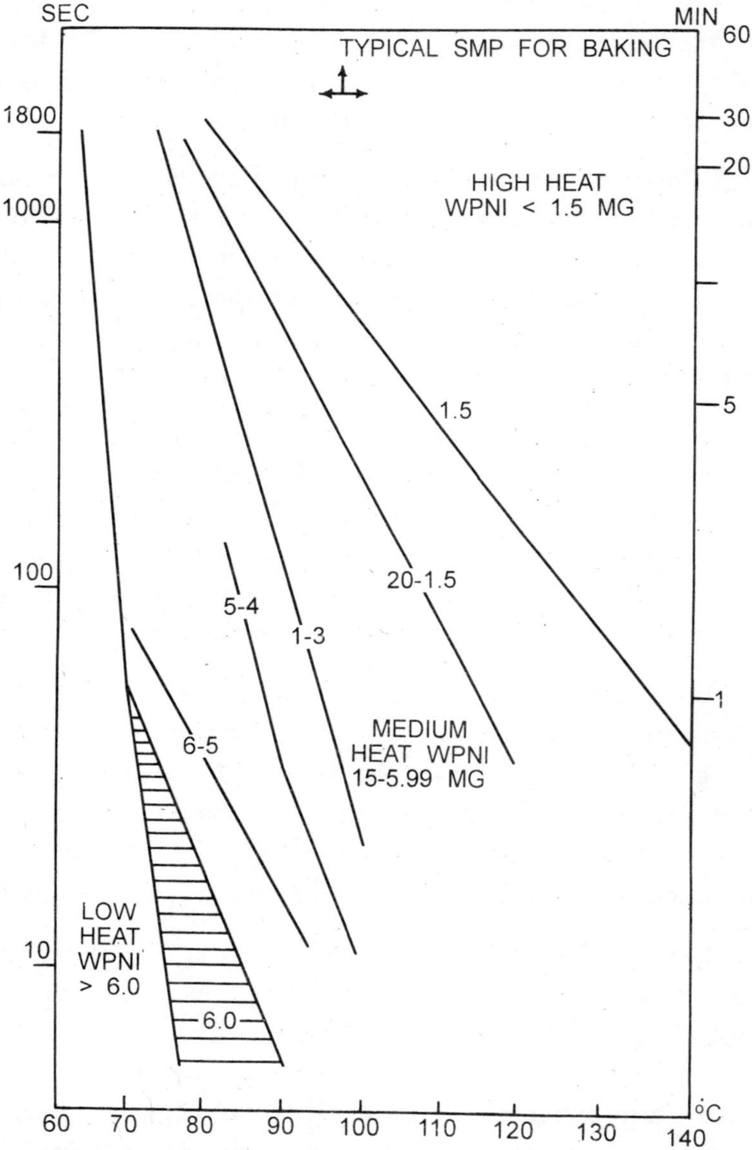

Fig. 4.14 WPNI (mg/g) of Powder as a Function of Pasteurization intensity, A relation between Temperature & time.

High heat powder was previously manufactured by heating the milk to 95°C and holding it for 20-30 minutes, but the large throughput that is common today makes this long holding impossible, as it would require a very large holding capacity. The long holding has been substituted by ultra high temperatures. Pre-heating equipment is required to heat the milk to temperature varying from 75 to 130°C and a holding capacity is required to effect holding from 15 seconds to 3-4 minutes which requires a detailed designing considering operation at 130°C for 20 hours in a day.

Heating of concentrate : Heating of concentrate to 75°C does not affect the level of denatured whey protein. For a given heat treatment, whey protein denaturation in Skim Milk Concentrate takes place in an inverse ratio to the degree of concentration.

Effect of Spray Drying : In general, practically no whey proteins denaturation takes place during spray drying and, therefore, the level of denaturation in dried milk is eventually equal to that of concentrate. Some denaturation may take place in a counter current dryer.

APPLICATION OF HEAT CLASSIFIED POWDER

A. Low Grade Heat Powder

In the manufacture of recombined cheese, where fresh milk can be stretched by addition of 30-40% reconstituted milk, it is important to use a low grade heat powder. Because excessive heat treatment of milk for use in the conventional manufacture of cottage cheese results in co-precipitation of the denatured whey proteins with acid casein curd and loss of desirable curd tension properties.

The organoleptic properties of the reconstituted milk are affected by the heat treatment of milk before drying. A cooked flavour is found in milk heated sufficiently to denature the whey proteins, as free sulphahydryl compounds are liberated.

Low heat powder is required for the manufacture of recombined pasteurized milk and ice-cream.

The low heat powder is of importance for recombining dairies to avoid a cooked flavour in the reconstituted milk. To avoid contamination of it from the water used for reconstitution, it is necessary to pasteurize the reconstituted milk.

B. Medium Grade Heat Powder

The medium heat powder is mainly used for sweetened condensed milk. when powder is used for the manufacture of condensed milk and evaporated milk, a very specific heat treatment is required to obtain the right viscosity and in the case of evaporated milk, the powder must be heat stable to sustain the sterilization. For these uses, the WPN is often not a sufficient criterion for the suitability of the powder. The right combination of temperature and holding time is found by testing the powder and varying the heat treatment until a satisfactory result is achieved.

C. High Grade Heat Powder

The high grade heat powder is used for manufacturing fermented products such as yoghurt and manufacture of cakes, biscuits etc. Milk processed for bread making must receive a sufficient heat treatment to inactivate unidentified factors responsible for gluten breakdown, extreme dough extensibility and poor loaf volume etc.

Whole Milk Powder

Whole milk is normally given a rather severe heat treatment before evaporation in order to develop the antioxidants that delay oxidation of the fat. Sulphahydryl compounds are released by the heat treatment and act as antioxidants. A more complete destruction of oxidizing enzyme also takes place.

Judging and Grading of Whole Milk Powder & Skim Milk Powder

A. Score Card

This is given jointly for both whole milk powder and skim milk powder in Table 4.4.

B. Procedure of Examination of Canned Product

(i) Sampling

Select a can of the product at random for examination.

(ii) Sequence of Observation

Cut more than three-fourths of the top of the can and turn it back. Then examine it in the following order :

(a) *Appearance of the can*

Look out for signs of rust etc., both outside and inside (when emptied).

Table 4.11

Score Card for Dry Milk (ADSA)

Items	Perfect Score
Flavour	45
Appearance	20
Package	5
Laboratory tests	30
Total Score	100

(b) *Flavour*

Note the smell (odour) immediately on opening the can. Reconstitute with distilled water to the original concentration (according to directions given on the can-label) and after about an hour examine the flavour (smell and taste) by tasting 5-10 ml on the tongue.

(c) *Appearance of the Product*

Examine uniformity of colour and absence of lumps, foreign specks, scorched/burnt particles etc.

(d) *Fineness and homogeneity (body and texture)*

Pour the product from the can into another container and note the flow characteristics.

Calculation

Problem 1

10,000 kg. milk testing 6.2% fat and 9% SNF is standardized for WMP manufacture, by removal of 60% fat cream. Its desired composition of WMP is 5% moisture and 26% fat, then calculate :

(i) Amount of cream to be separated.

(ii) Amount of WMP made from standardized millk.

Solution

Suppose, x kg. of cream to be separated from milk and y kg. of WMP to be manufactured.

$\because$ 100 kg milk contains 6.2 kg. fat.

$\therefore$ 1 kg. milk contains $\frac{6.2}{100}$ kg. fat

10,000 kg milk contains $\frac{6.2}{100} \times 10{,}000 = 620$ kg. fat.

$\because$ 100 kg. cream contains 60 kg. fat.

$\therefore$ 1 kg. cream contains $\frac{60}{100}$ kg. fat.

x kg. cream contains $\frac{60}{100} \times x = 0.6x$ kg. fat

100 kg WMP contains 26 kg. fat.

$\therefore$ 1 kg WMP contains $\frac{26}{100}$ kg. fat.

y kg WMP contains $\frac{26}{100} \times y = 0.26\ y$ kg. fat.

By fat equation,

$0.6x + 0.26y = 620 \quad \ldots \quad$ (i)

$\because$ % SNF in cream

$\frac{100 - \text{fat\% in cream}}{100 - \text{fat \% in milk}} \times$ % SNF in milk.

$= \frac{100 - 60}{100 - 6.2} \times 9 = \frac{40}{93.8} \times 9 = 3.837$ or 3.84% SNF

% SNF in cream = 3.84%

100 kg. cream contains 3.84 kg. SNF

$\therefore$ 1 kg. cream contains $\frac{3.84}{100}$

x kg. cream contains 0.0384x kg SNF

% SNF in WMP = 100 (Moisture % + fat%) in WMP

$= 100 - (5 + 26)$

$= 69$

$\therefore$ % SNF in WMP = 69%

100 kg. WMP contains 69 kg. SNF

$\therefore$ 1 kg. WMP contains $\frac{69}{100}$ kg. SNF

$\therefore$ y kg. WMP contains $\frac{69}{100}$ y = 0.69y kg. SNF

100 kg. milk contains 9.0 kg SNF

$\therefore$ 1 kg. WMP contains $\frac{9}{100}$ kg. SNF

10,000 kg. WMP contains $\frac{9 \times 10{,}000}{100}$ = 900 kg. SNF

Total SNF in 10,000 kg. milk by equation

$$0.0384x + 0.69y = 900 \quad \ldots \quad (2)$$

From equation (1) × 0.0384 – (2) × 0.6 then we get

$$0.6x \times 0.0384 + 0.26y \times 0.0384$$

$$= 630 \times 0.0384 \quad \ldots \quad (3)$$

$$0.0384x \times 0.6 + 0.69y \times 0.6$$

$$= 900 \times 0.6 \quad \ldots \quad (4)$$

(3)-(4) $\quad 0.26y \times 0.0384 - 0.69y \times 0.6$

$$= 630 \times 0.0384 - 900 \times 0.6$$

$$\therefore 0.009984y - 0.414y = 24.192 - 540.0$$

$$- 0.404016y = -515.808$$

$$\therefore y = \frac{515.808}{0.404016} = 1276.701913 \text{ or } 1276.70$$

$\therefore$ y = 1276.70 kg.

Putting the value of y in equation (1) we get

$$0.6x + 0.26y = 630$$

$\therefore$ 0.6x + 0.26 × 1276.70 = 630

$\therefore$ 0.6x + 331.942 = 630

0.6x = 630 − 331.942

0.6x = 298.058

$$\therefore x = \frac{298.058}{0.6} = 496.7633 \text{ or } 496.76$$

$\therefore$ x = 496.76 kg.

Hence,

(i) The amount of cream to be separated 496.76 kg.

(ii) Amount of WMP manufactured 1276.70 kg.

Problem 2

8,000 kg. of milk testing 6% fat and 9% SNF is standardized for WMP manufacture by addition of Skim milk 0.1% fat and 8.8% SNF. If the desired composition of WMP is 4% moisture and 25% fat.

(i) Amount of skim milk required

(ii) Amount of whole milk required.

Solution

Suppose, x kg. of skim milk to be added and y kg. WMP be made.

$\because$ 100 kg. milk contains 6 kg. fat.

1 kg. milk contains $\frac{6}{100}$ kg. fat

8,000 kg. milk contains $\frac{6 \times 8,000}{100} = 480$ kgs. fat

$\because$ 100 kg. milk contains 9 kg. SNF

1 kg. milk contains $\frac{9}{100}$

8,000 kg. milk contains $\frac{9}{100} \times 8,000 = 720$ kg. SNF

$\therefore$ 100 kg. of skim milk contains 0.1 kg. fat.

1 kg. milk contains $\frac{0.1}{100}$

x kg. milk contains $\frac{0.1}{100}$ x = 0.001x kg. fat.

$\therefore$ 100 kg. WMP contains 25 kg. fat

$\therefore$ 1 kg. WMP contains $\frac{25}{100}$ kg. fat

y kg. WMP contains $\frac{25}{100}$ y or 0.25 y kg. fat.

$\because$ 100 kg. skim milk contains 8.8 kg. SNF.

$\therefore$ 1 kg. skim milk contains $\frac{8.8}{100}$ kg. fat

x kg. skim milk contains $\frac{8.8}{100} \times$ x or 0.088x kg. SNF

SNF % in WMP

100 – (Moisture % + fat %) in WMP

$$= 100 - (4 + 25)$$

$$= 100 - 29 = 71\%$$

$\because$ 100 kg. WMP contains 71 kg. SNF.

$\therefore$ 1 kg. WMP contains $\frac{71}{100}$ kg. SNF

y kg. WMP contains $\frac{71}{100}$ y = 0.71y kg. SNF.

$$480 + 0.001x = 0.25y$$

or $0.25y - 0.001x = 480$ ... (1)

$$720 + 0.088x = 0.71y$$

or $0.71y - 0.088x = 720$ ... (2)

equation (1) × 0.088 – equation (2) × 0.001 then we get.

$$0.25y \times 0.088 - 0.001x \times 0.088$$

$$= 480 \times 0.088 \quad \ldots \quad (3)$$

$$0.71y \times 0.001 - 0.088x \times 0.001$$

$$= 720 \times 0.001 \quad \ldots \quad (4)$$

(3) – (4) $0.022y - 0.00071y = 42.24 - 0.720$

or $0.02129\ y = 41.52$

or $y = \frac{41.52}{0.02129}$

$$y = 1950.21$$

Putting the value of y in equation we get

$$0.25y - 0.001x = 480$$

or $0.25 \times 1950.21 - 0.001x = 480$

or $487.5525 - 480 = 0.001x$

$$\therefore x = \frac{487.5525 - 480}{0.001} = \frac{7.5525}{0.001}$$

$$\therefore x = 7552.5$$

Hence,

The amount of skim milk required 7552.5 kg.

The amount of whole milk powder made 1950.21 kg.

Reconstitution and Recombination of Milk

Reconstituted milk is the liquid milk obtained by adding water to skim milk powder (SMP) or whole milk powder (WMP).

Recombined milk is the liquid milk obtained by adding water to SMP and adding milk fat separately in such a quantity that the desired fat content is achieved.

Basic Reconstitution and Recombining Operation

1. Raw Material Handling

(a) Milk Powder : Non-fat solids for recombined milk are usually supplied in the form of skim milk powder.

The powder can be stored for months or even for years, without getting spoiled, and dissolved easily in water to form reconstituted milk. The dissolving properties of milk powders are very important for recombined product quality.

Factors affecting the dissolving of milk powder are :

(i) *Wettability* : Agglomerated powder has improved capillarity, resulting in increased wettability. Increased particle size (130-150 mm) also results in improved wettability. Good wettability is less than 30 seconds.

(ii) *Solubility* : This property describe how well the powder dissolve and form a stable suspension. The solubility depends very much on the technology used for production of the powder.

A good solubility index should be as low as 0.25 ml undissolved sediment in 50 ml recombined milk.

(iii) *Ability to sink* : The ability to sink is a function of specific volume and particle size. Agglomerated powder normally has the best ability to sink.

(iv) *Dispersability* : Good dispersability is obtained when powder added to the water is distributed as single particles, leaving no lumps. The structure of powder particles, as well as the configuration of the protein molecules is important for this property. A powder with high content of denatured proteins is very difficult to disperse. A dispersability of at least 90% is normal for milk powder for recombination.

2. Fats and Oil

Unsalted butter can be used in the manufacture of recombined milk products, but it must be kept under refrigerated storage.

The most common source of milk fat for recombination is anhydrous milk fat (AMF), which does not require such storage. It is typically packed in cans and drums. The AMF is packed under nitrogen gas (inert gas) packaging. AMF can be kept for 6-12 months even at elevated ambient temperatures of 30-40°C.

Milk fat packed in cans can be melted by immersion in hot water at 80°C for 2-3 hours. Drums of AMF, however, require longer melting times. The normal method is to store the drums in a hot room at 45-50°C for 24-28 hours before use. Once melted, the AMF should be transferred to a jacketed holding tank with facilities for maintaining the temperature.

Similar handling system can also be employed when non liquid vegetable oils are used in production of recombined "filled" milk products.

3. Water

Water is one of the raw materials of all types of reconstituted and recombined milk. It must be of good drinking quality free from harmful micro-organisms and of acceptably low hardness. Too much copper and iron in the water may cause off-flavours, due to oxidation of fat. The maximum levels recommended are Cu (0.5 mg/l) and Fe 0.1 mg/L.

4. Additives

Dry additives as sugar, stabilizers and emulsifiers can be handled in the same way as the milk powder, i.e., they are dumped from the bags either directly into mixing vessel or into the mixing system.

RECOMBINATION OF MILK POWDER

Small Scale Production : In small scale production 1000-2000 L, batch processing is most common. Mixing and processing are carried out in a jacketed mixing tank with a two-speed shearing agitator and with heating and cooling facilities. (Fig. 4.15)

Large Scale Production : Above 2000 Processing (Fig. 4.16)

Temperature and Hydration Time

(i) The solubility of the powder increases when the water temperature increases from 10 to 50°C.

(ii) Low-heat powder is easier to disclose than high-heat powder.

(iii) It is important that the proteins obtain their normal state of hydration, which takes less than 20 minutes at 40-50°C.

(iv) Recombined milk to cheese manufacture should be given two hours' hydration time.

(v) The fat must be added at a temperature above its melting point. AMF must be added at above 40°.

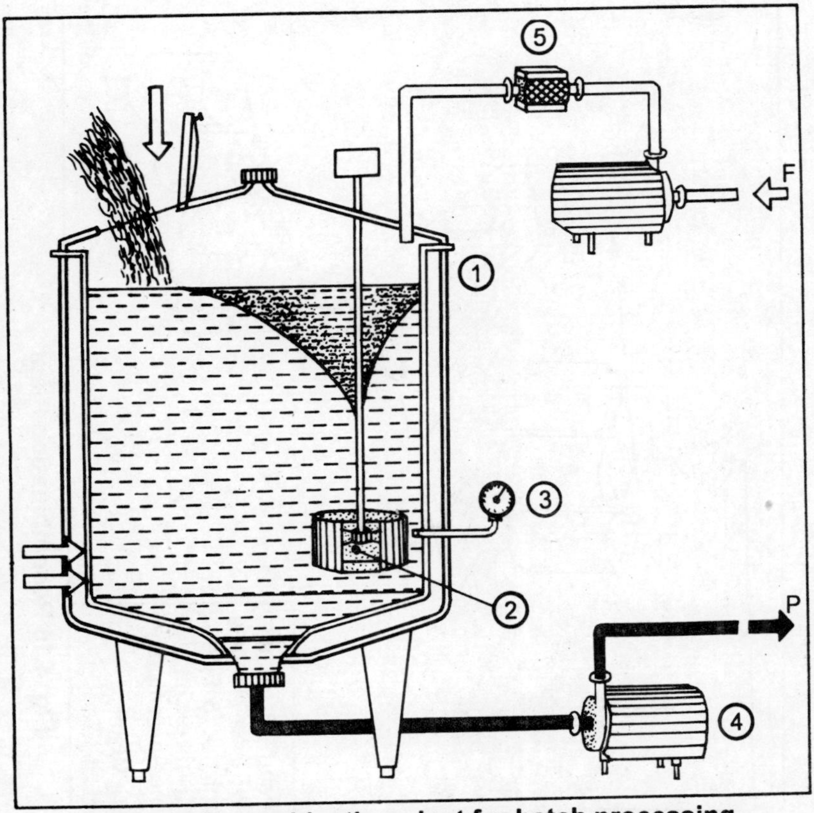

Fig. 4.15 Recombination plant for batch processing

1. Mixing tank with heating/cooling jacket.
2. High-Shear two speed agitator.
3. Thermometer.
4. Emptying pump
5. Flow meter

MP = Milk Powder, F = Fat, P = Recombined milk.

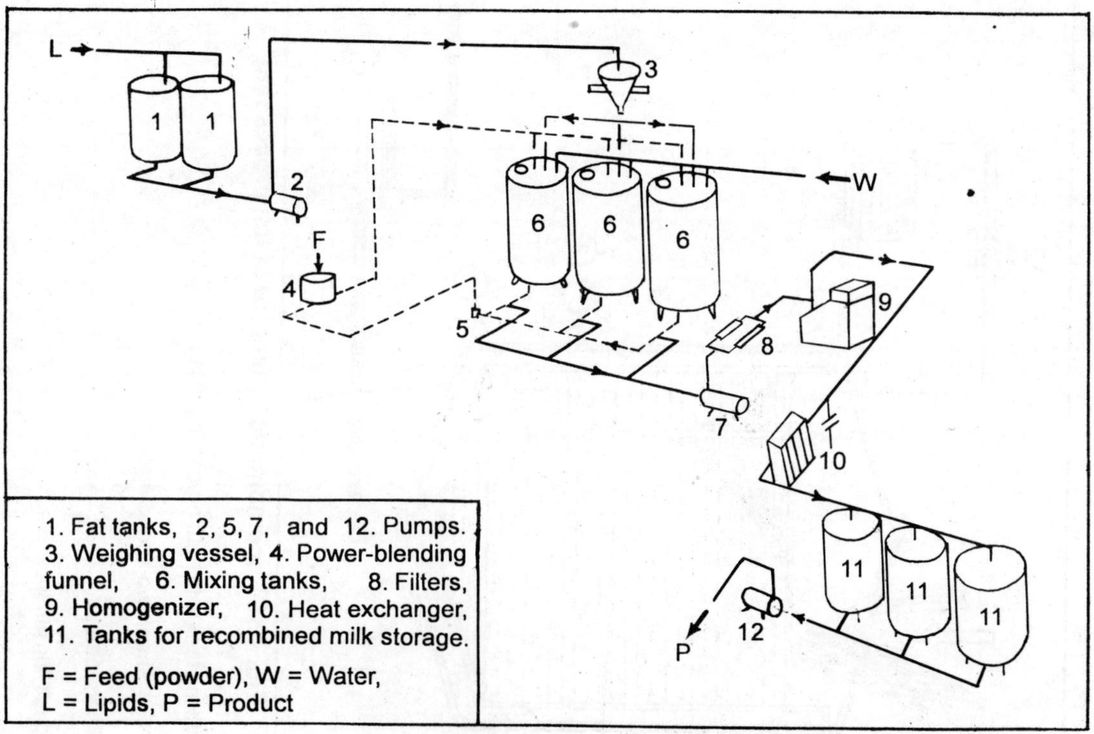

Fig. 4.16 Recombination Plant

Courtesy : Alfa-Laval.

Fat Addition and Emulsification

(i) Fat addition into recombined products has been a relatively different process. Traditionally, the melted fat is metered into the line during continuous operation, followed by through mixing in a static or mechanically operated mixer before entering a homogenizer.

(ii) In modern system, using high-shear mixing devices, fat can be dosed directly into the mixing vessel of the unit. The dispersion of fat is sufficient to create a stable emulsion which allows eliminating homogenization at this stage.

(iii) An emulsifier is sometimes added to faciliate and improve the emulsification of milk fat.

Air Content

(i) Test indicates that the air content in reconstituted skim milk, dissolved at 50°C and with 14 to 18% dry solids, is the same as in normal skim milk.

(ii) The mixing equipment may also cause air addition if not properly maintained. The air content of reconstituted milk increases at lower mixing temperature.

(iii) As recombination is accompanied by foaming, the volume of the mixing tanks' should be about 20% larger than the volume of the batch, to avoid foam forcing its way out of the manhole.

(iv) Some mixing equipments make it possible to recombined products under vacuum. By constantly extracting air from the mixing vessel, a certain vacuum is maintained, which reduces foaming to a minimum.

Disadvantages of Too Much Air in the Recombined Milk

- Foaming
- Burning-on in the pasteurizer.
- Cavitation in the homogenizer.
- Whey formation in cultured-milk products.
- Increased risk of oxidation of fat.

Heat Treatment

Three alternative methods are used :

(i) Pasteurisation at a temperature of at least 72°C for 15 seconds followed immediately by cooling to 4°C.

(ii) In-container sterilization of milk for 30 to 45 minutes at around 110°C followed by cooling to 38-54°C in the sterilizer.

(iii) UHT treatment by direct or indirect heating to 132-149°C for a few seconds, followed by cooling to approximately 20°C prior to aseptic packing.

PACKING AND STORAGE

The milk should be packed as soon as possible after production.

(i) Pasteurized milk can be packed in paper or plastic packs or glass bottles. If bottles are used, they should be a dark glass, which prevents the flavour of milk from being spoiled by exposure to light. Pasteurized milk must be kept in cold storage room.

(ii) UHT-treated milk must flow in closed aseptic system to the aseptic carton or canfilling machine.

Precautions/Notes

Insufficient hydration time may lead to a 'chalky' defect in the final product.

DISCUSSION QUESTIONS

1. Define milk powder. Explain its physiochemical attributes.
2. What is the role of milk constituents in powder manufacture ?
3. Explain the technology of manufacturing whole milk powder.
4. Define roller drying. Explain the factors that affect the roller dried milk powder.
5. Explain the principles of spray dried milk powder.
6. What do you understand by Instantization ? How is WMP instantized ?

7. Narrate the manufacturing of skim milk powder.
8. Explain the relationship between reconstitutability and physical structure of powder.
9. Define fortification ? How is it done ?
10. What do you understand by heat classified milk powder ?
11. Explain Whey protein nitrogen index.
12. How will you judge and grade milk powder ?
13. Distinguish between reconstitution and recombination of milk powder.
14. Explain packaging and storage of milk powder.

5

Dryers and Accessories

Drying is an age-old practice which has been indicated even through the accounts of famous explorer Marco Polo. He traced out that Mongolians in Asia used to make milk powder by drying milk in the sun. Those were ancient practices of drying which may not sound good in terms of accuracy and precision but, today, it is a hightech processing technology that has gone through a series of modernisation and modifications. From Roller dryer to Spray dryer there is a mammoth change in the equipment as well as in processing technology, that are running rampant all over the globe. Spray dried skim milk powder is the most coveted milk product today in terms of its shelf-life (upto three years) and has a tremendous potential for reconstitution.

ROLLER DRYER

Principles of Roller Drying

Roller drying can only be used for liquid products which can either have a low viscosity or be highly viscous to paste like. Evaporation is the main process in roller drying. The metal rollers are heated from inside with condensing steam and the product dries as a thin film by contact with the cylinder surface. So, it is an indirect type of dryer.

Application of Roller Dryer

Roller dried skimmed milk powder is used mainly for the production of animal feeding stuff and mixed feed. Roller dried whole milk is normally used in milk chocolate manufacture.

Operating Characteristics of Roller Dryer

- The milk is spread as a thin film on the surface of the hot drum and after one revolution is scraped off by a knife in the form of flakes/ scales or powder. The knife is sharp, hard and

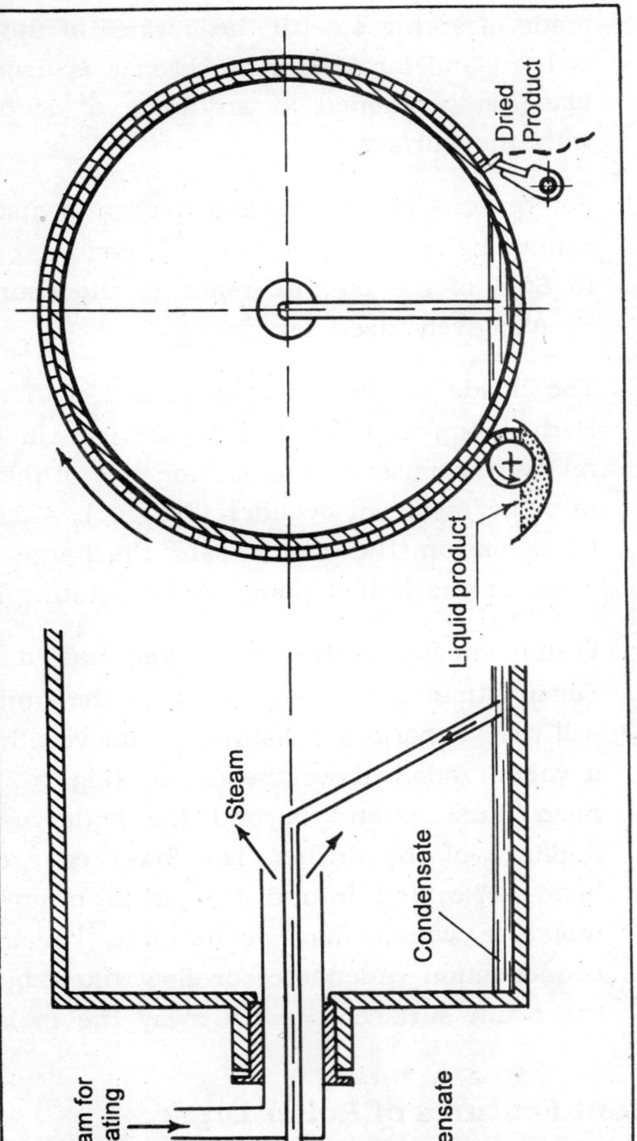

Fig. 5.1 Diagram of the steam inlet into the drum for heating and the condensate discharger in roller dryers

flexible, known as 'doctor blade'. The blade made of spring steel if the surface of the drun is hard and for soft drum, bronze is used. Th blade is positioned at an angle of 15 to 30 with the surface.

- For reasons of construction, feed and discharge cannot be at the same point. Therefore, only 70 to 80% of the circumference of the drum car be effectively used for drying.
- The inside of the drum is heated with steam Both steam and condensate are led via sealed rotary transmission through one face of the roller into the inside of cylinder (Fig. 5.1). After heat transmission the condensate discharge takes place at the lowest point of the rotating drum.
- Water vapour above the dryer has a lower density than the air surrounding the unit, and will rise. Vaporized moisture is removed through a vapour hood above the drums (Fig. 5.2). The hood must extend beyond the ends and the midlines of the drums. The lower edge of the hood is formed into a trough to drain away moisture which may accumulate because of condensation. Adequate air flow must be over the drum surface to carry away the moisture.

Important Features of Roller Dryer

Drum diameter : 500 to 1500 mm.

Drum length : 1000 to 3000 mm.

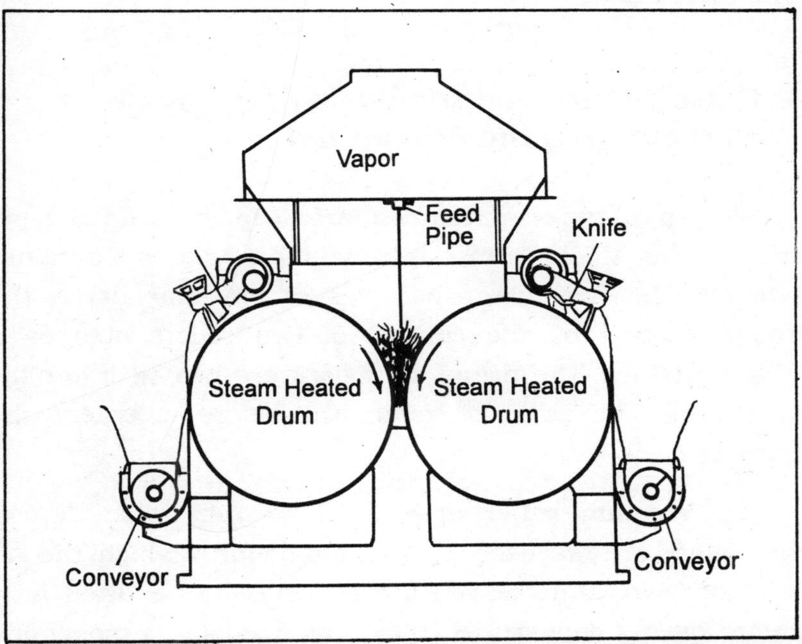

Fig. 5.2 Diagram of Roller Dryer (Courtesy : Blow-Knox Company)

Thickness of Drumwall : 20 to 40 mm.

Thickness of the film of the product : 0.1 to 0.5 mm.

Rate of revolution : 5 to 30 revolution/minute.

Steam temperature : 120-165°C

Steam pressure : 2 to 7 bar.

Specific Heat Consumption : 3000-3500 KJ/kg. water removed

Specific Steam Consumption : 1.3-1.5 kg. Steam/kg. water removed

Specific Water evaporation : 10-30 kg. water/m^2.h
(in special cases, upto 60 kg. water/m^2)

Classification

A. *Classification on the Basis of Pressure Surrounding the Production*

1. Open type or Atmospheric type : In open type, one or two steel drums or rollers dip into a container which holds the milk to be dried or on some dryer, the trough formed by the contact of two rollers acts as a milk container. The drums or rollers are heated internally by steam. The milk is condensed before it enters the dryer (Fig. 5.3).

2. Vacuum roller dryer : In vacuum roller dryers, the rollers are enclosed in a chamber form which the air is withdrawn, thus permitting the milk to be dried at a much lower temperature than that used for atmospheric drying (Fig. 5.4). Drying of the milk at a temperature below 100°C at 91-98 kPa means that the effects of oxygen and high temperature markedly reduced. The product has better properties than the milk powder obtained from roller drying under atmospheric pressure.

B. *Classification on the Basis of Number of Drums*

There are single and double roller dryers

(i) Single Roller Dryers

In a single roller dryer, drying occurs on only one heated roller (Fig. 5.5). The main difference lie, in the single roller dryer, in the way the product is applied.

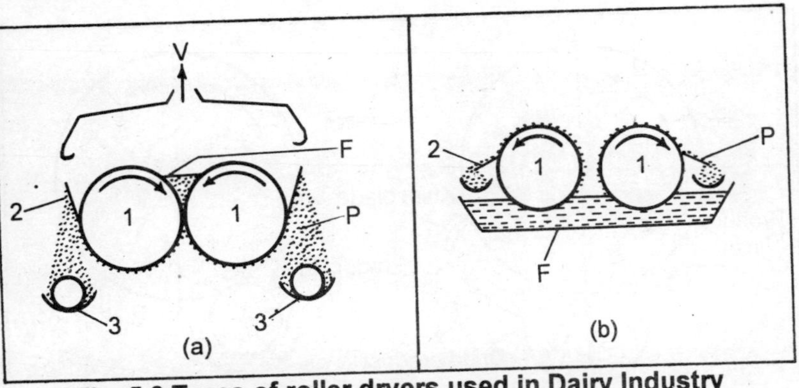

Fig. 5.3 Types of roller dryers used in Dairy Industry

(a) Double drum with top feed. (b) Twin drum with dip feed

1. Drum, 2. Knife, 3. Conveyors, F = Feed. V = Vapour, P = Product.

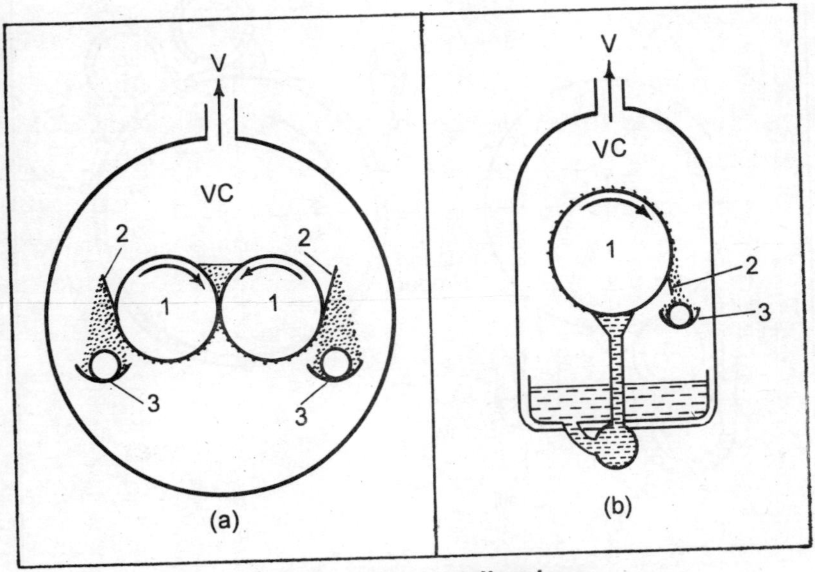

Fig. 5.4 Vacuum roller dryer

(a) Vacuum double drum with top feed
(b) Vacuum Single drum (Pan feed)

1. Drum, 2. Knife, 3. Conveyors, 4. Pump,
F = Feed. V = Vapour, Y_c = Vacuum P = Product

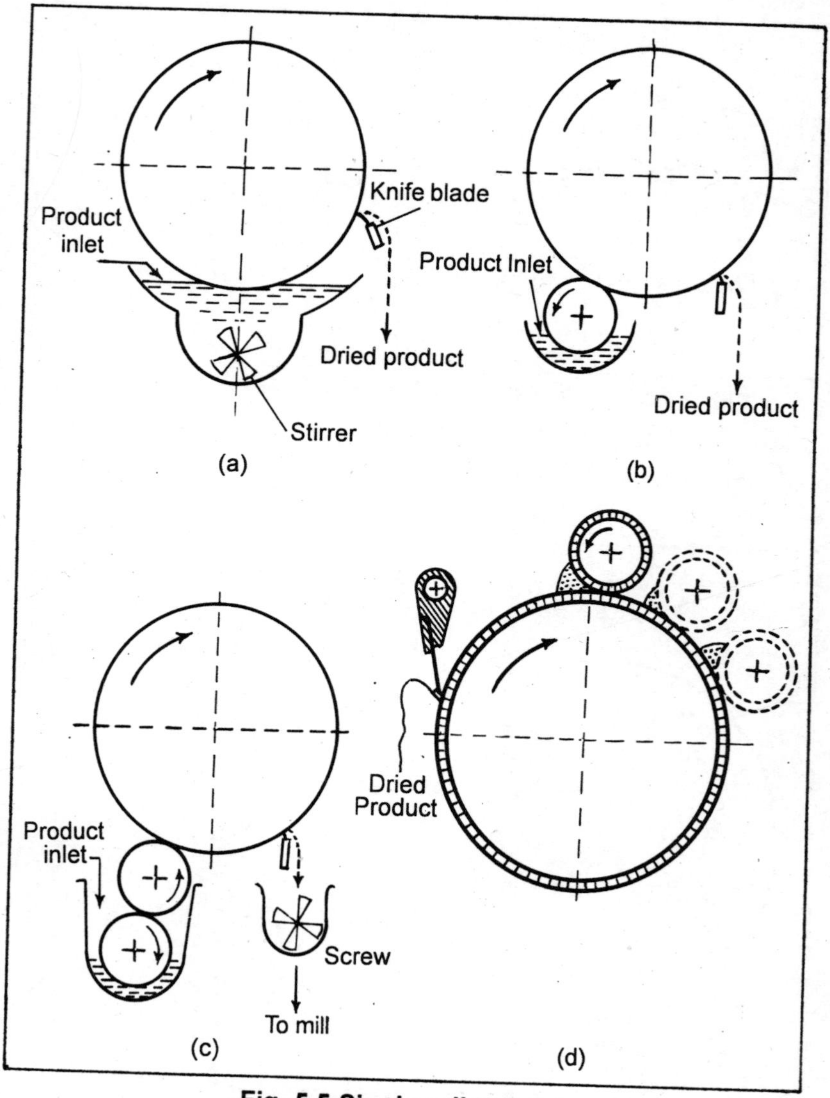

Fig. 5.5 Single roller dryer

(a) With Dip feed
(b) Application by feed roller
(c) Thin layer application with dip roll and transfer roller
(d) Feed rollers on top

Source : Kessler 1981.

(a) *Single roller dryer with dip feed*

In this type, the roller dips into the product. (Fig. 5.5a). The product is mixed by a stirrer to even out temperature and concentrations. The disadvantage of this method is that the drum heats the product in the trough and that the thickness of the film is not uniform.

(b) *Single roller dryer with feed application by feed roller and thin layer application with dip roll and transfer roller.*

These both types of feed applications are shown in Fig. 5.5b & 5.5C, which are often used in dairy industry which allows a uniform thin film to the drum.

(c) *Single roller dryers with feed roller on top*

Fig. 5.5d shows this type of roller.

(ii) Double or Twin Roller Dryer

In twin roller dryer, drying occur on double heated roller or drum. (Fig. 5.6 & Fig. 5.7)

(a) *Twin roller dryer with top feed*

In top feed dryer, where the product is supplied to the drums by a reservoir. The drums draw the product into the clearance between them and a film of product

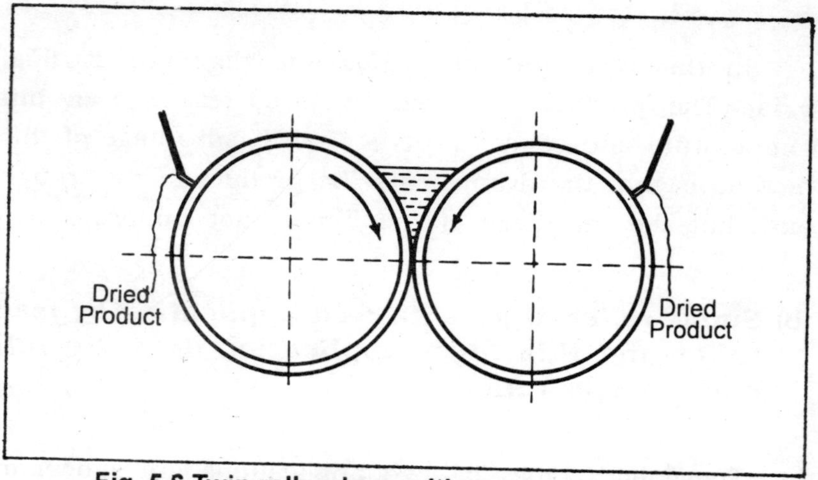

Fig. 5.6 Twin roller dryer with reservoir (top feed)

Source : Kessler (1981)

adheres to each other (Fig. 5.6) The clearance between the rollers is adjustable by one movable and one immovable drum. The high temperature of the product in the reservoir is a disadvantage since concentration by evaporation takes place prematurely. The quality of heat sensitive milk products can be impaired.

(b) *Twin roller dryer with dip feed*

Fig. 5.3b shows the twin roller dryer with dip feed. Deep feed is used for certain suspensions of solids, usually with recirculation of material in the tray.

(c) *The twin roller dryer with spray film feed*

It has been used for many products in the dairy Industry (Fig. 5.7) Many small discs driven by a shaft rotate and dip into the liquid in a distributing reservoir. Due to centrifugal force a liquid bulge is formed on the disc's circumferences, which is removed at a selected place by a steam of air from nozzles (Fig. 5.7b, 5.7c) and sprayed on surface of the drum.

SPRAY DRYERS

The Spray drying process, in recent years, has become the acceptable method of drying the milk. In India, most of the milk drying plants are spray driers.

Principle of Spray Drying

The principle in spray drying is that the milk is atomized (converted into a fog like mist) into a drying chamber, through which passes a stream of hot air, thereby the milk dry to powder particles, which fall towards the bottom of the chamber from where it is removed more or less continuously in various ways.

Air inlet temperature upto 215°C are used for the drying of milk, but due to the evaporation, the temperature drops immediately in the drying chamber and the milk solid will not reach a temperature approaching the air inlet temperature. The temperature in the drying chamber is equal to the air outlet temperature, which is approximately 95°C when single stage drying is used and

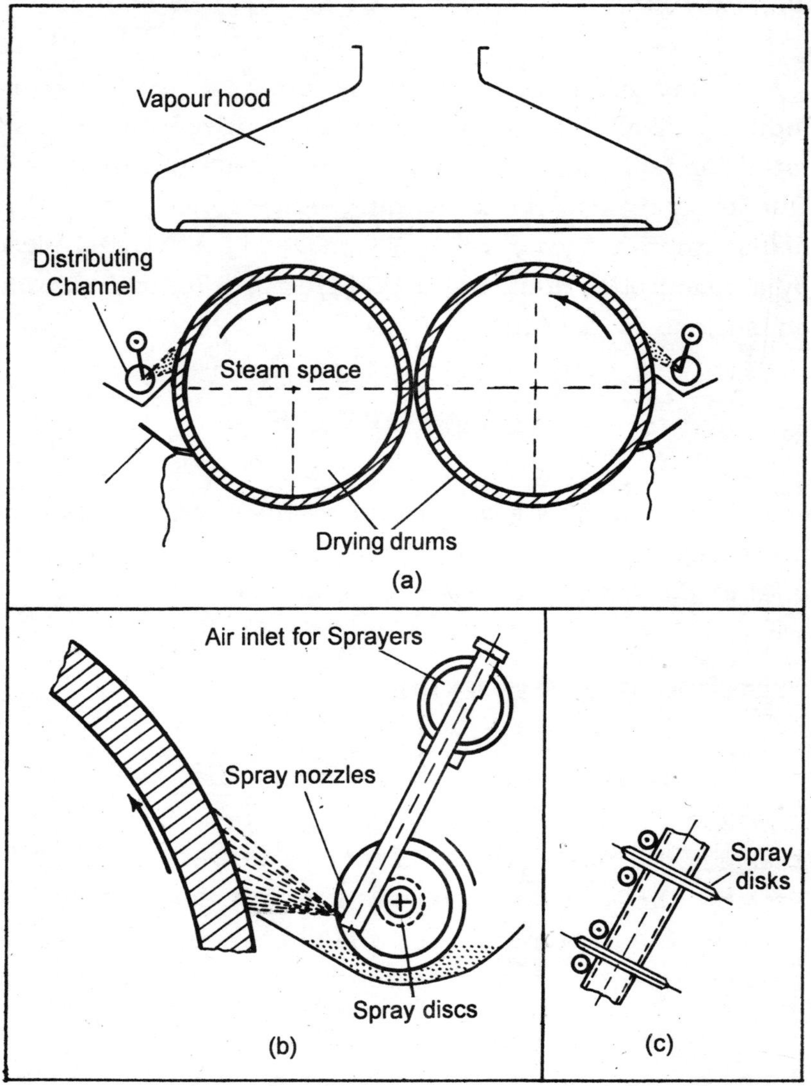

Fig. 5.7 (a) Twin roller dryer with spray film feed, (b) Spraying portion, (c) Spray disc

Source : Kessler (1981)

lower with multi-stage drying. The product temperature will be 20-30°C below the air outlet temperature provided a concurrent dryer (i.e., a dryer where product and air move in the same direction) is used.

Advantages of Spray Drying

- Drying period is very short
- High production rate, with superior flavour, appearance and solubility of product.
- Gentle drying method.
- Continuous single-step operation.
- Uniform product and no product contamination.
- Low dryng cost.
- High thermal efficiencies.
- Heat contact to product is short during drying and removal.

Disadvantages of Spray Drying

- The initial cost is high and the process is only economical when large quantities of milk are handled.
- Spray dried plant requires a tall building to house it.
- Bulk density of the dried particles is often difficult to keep constant.

COMPONENTS OF SPRAY DRYER

Air Supply System

The ambient air is filtered, heated by steam or liquid phase heating system with oil or gas upto the temperature of 150-250°C and introduced into the drying chamber at a velocity upto 50m/second.

(a) *Air filtration* : Air filters are essential to avoid contamination from the air of the powder. Air cleaning is usually performed by dry filters that are cleanable or disposable. Presently, filter is supplied automatically from one roll and collected on another roll after use. The filter for drying air is not required .to destroy possible airborne bacteria because the bacteria will not survive when the air is heated. Since air pollution continues to increase, it is necessary to introduce more rigorous air purification for spray drying of milk and milk products—Air is usually filtered before it is heated.

(b) *Air heating Systems* : For drying milk, air is used at temperature usually from 150-250°C. Air is either indirectly heated in steam air heaters or in oil or gas-fitted air heaters or directly heated by means of gas or oil.

Steam Air-heater

It usually consists of a system of fined copper tubes. It may also be made of glavanized steel or stainless steel. High pressure steam alone, or high-pressure steam combined with low pressure steam, is used. The steam heats the air to about 10°C below the temperature produced by steam heater. Table 5.1 shows the saturated vapour temperature corresponding to each vapour pressure. By combining the steam air-heater with an electric air heater or gas burner, higher temperature can be obtained.

Table 5.1

Saturated Vapour Temperature Corresponding to Each Vapour Pressure

Vapour pressure	Vapour temperature
10 bar	180°C
15 bar	198°C
20 bar	212°C
25 bar	224°C
40 bar	250°C

Source : Kessler, 1981

Oil or Gas Fired Air Heaters

Higher air temperatures are obtained by the use of oil or gas-fired air heaters. A better heat economy is achieved by the process. Approximately 14 kg. of oil is required for each 100 kg. of water evaporated in the dryer with an air temperature of 200°C. The air heater consists usually of concentric cylinders. Air can be heated upto 340°C.

These two types of air heaters are commonly used.

Direct Heating System

In this system, the product of combustion of gas or light oil is mixed with a suitable amount of atmospheric air. The thermal efficiency of such an air heater is practically 100 per cent. It has not been used for drying milk in India.

(c) Air Blowers

The movement of air through the dryer is carried out by centrifugal fans or blowers. One or two main fans are used. For dryers that operate with a single exhaust fan, the drying chamber is under a comparatively high vacuum. Most of the spray dryer, however, operate with a system using two main fans. Apart from the exhaust fan, a supply fan is mounted on the inlet side of the drying chamber. This fan sucks air through an air filter and force it through the air heater and dispenser. Two fan system provides for great flexibility of chamber pressure.

Air Flow

Due to high temperature, the air entering the spraying chambers has very low humidity. Hot air distribution is very important after it has been inducted into the chamber. It should have streamlined (laminar) flow, i.e., the chamber must not have dead corners (Fig. 5.8).

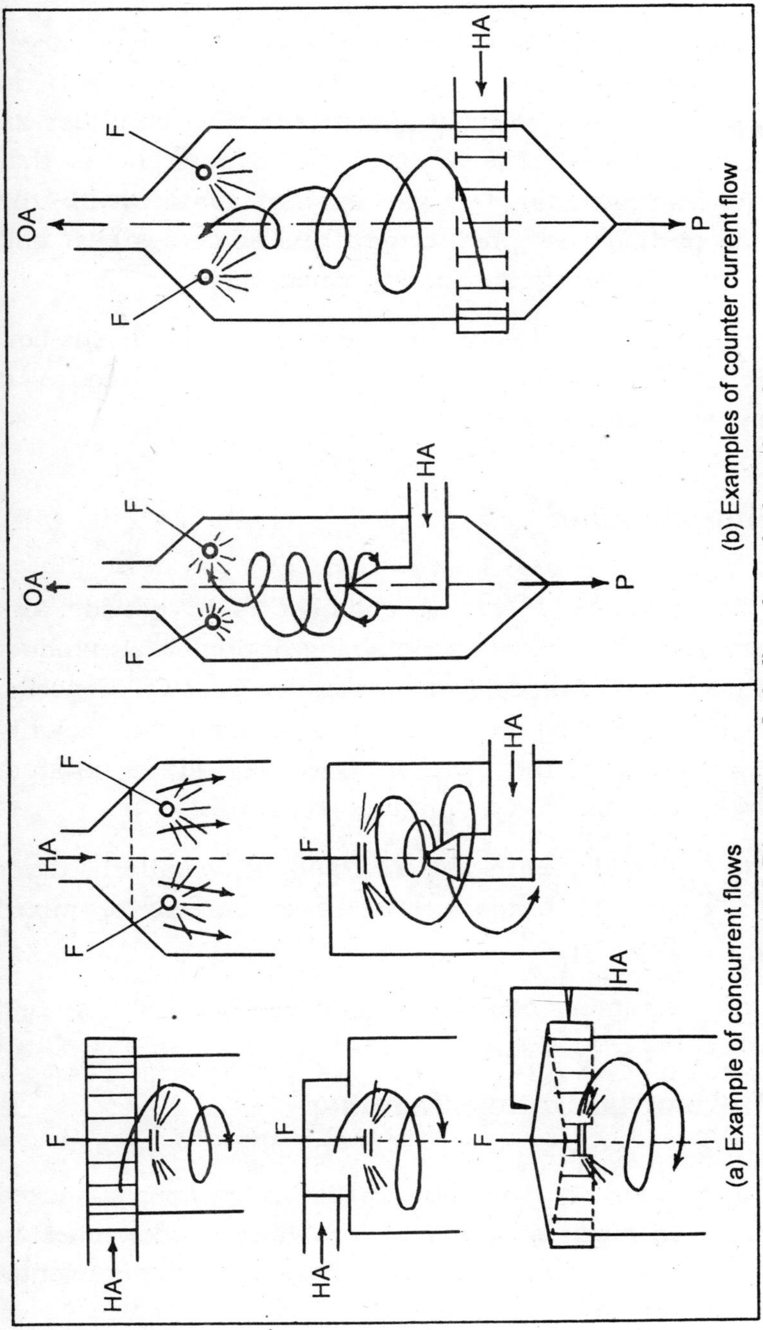

Fig. 5.8 Direction of air flow in a drying chamber

HA = Hot air, P = Product, F = Feed, OA = Outlet air

The air stream moves either through the spray dryer in the same direction as milk flow (concurrent flow), in the opposite direction (counter current flow) or under an angle (mixed flow). The advantage of counterflow is that the hottest (incoming) air comes into contact with the already partially dry product and this enhances heat and mass transfer, reducing energy consumption.

In spite of the inferior heat economy, concurrent flow predominantly used in the dairy Industry because it enhances product solubility.

Drying Chamber

The function of drying chamber is to provide air/ particle residence times for obtaining desired dried product moisture levels without heat degradation and wall deposits. Dryer chambers may be designed to discharge the majority of the powder at the base or to convey all the product with the exhaust air to product separator.

(i) Based on the air and milk flow pattern, dryer can be concurrent, counter current or mixed flow type.

(ii) Chamber can be Horizontal or Vertical (Fig. 5.9)

(a) Horizontal Drying Chamber

In this type of dryer (Fig. 5.10). The powder is sucked from the chamber by means of a rotating powder collector. These are box type. The advantages of the horizontal chamber are :

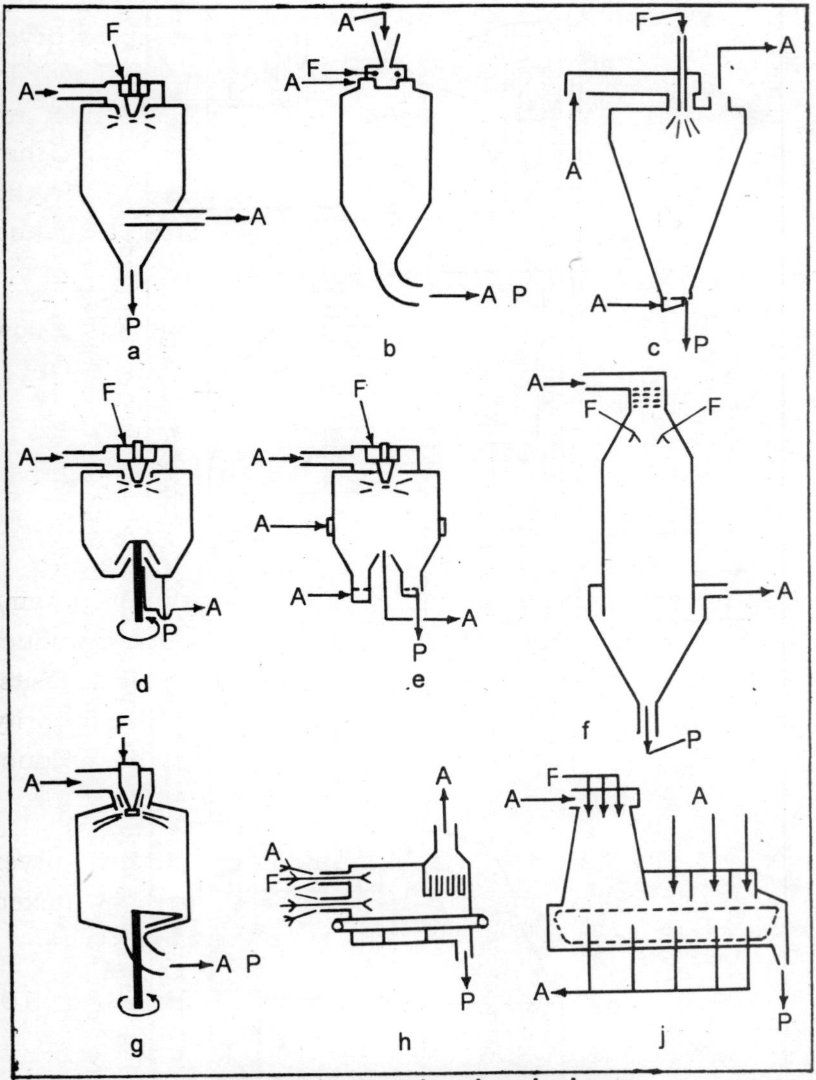

Fig. 5.9 Drying chamber designs

(a) Conical-based chamber with two-point product discharge, (b) Conical-based chamber with single-point product discharge, (c) Conical-based chamber with integrated static fluid bed, (d) Inverted base coned chamber, (e) Inverted base coned chamber with integrated static fluid bed, (f) Tower-form (nozzle tower) chamber, (g) Flat-based chamber with product sweeper, (h) Box chamber with integrated screw conveyor, (i) Box chamber with integrated conceying band. F = Feed, A = Air flow, P = Product

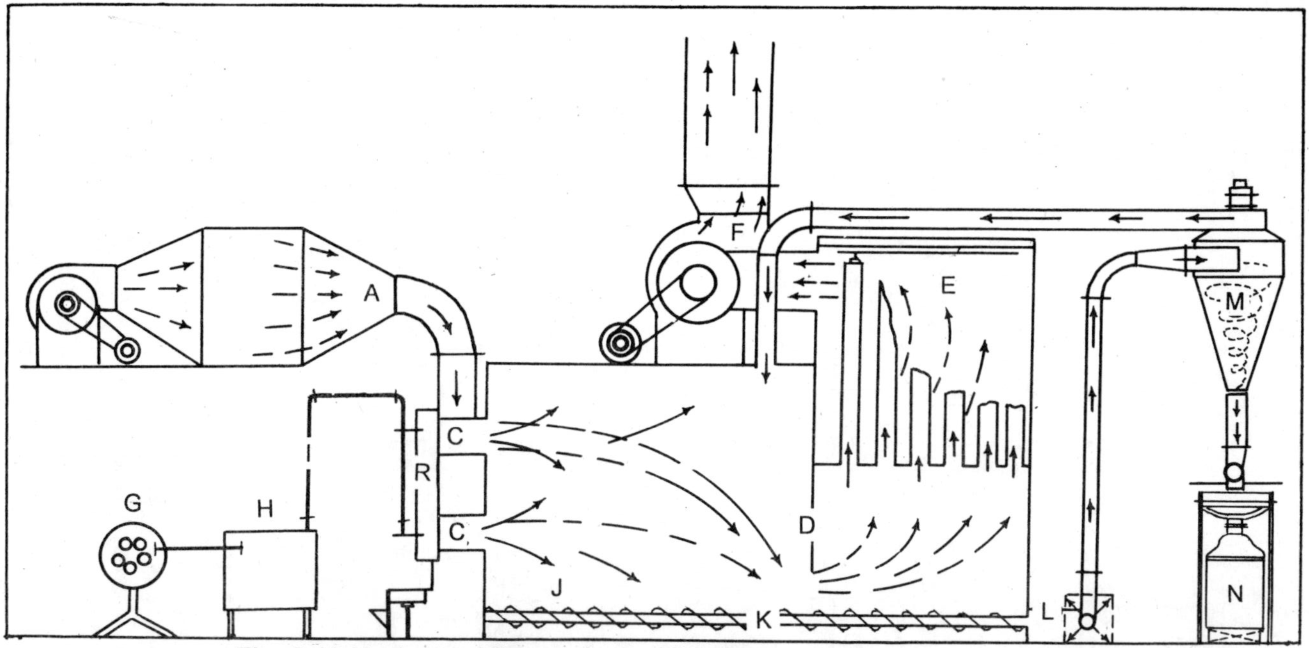

Fig. 5.10 Horizontal Spray dryer. (***Courtesy*** **:** C.E. Rogers Company)

A. Air Supply, B. Cone distributor, C. Air inlet and spray nozzle, D. Baffle, E. Cloth filterbag, F. Air exhaust. G. Cone preheater, H. High pressure line, J. Reciprocating conveyor, K. Screw Conveyor, L. Outlet chamber, M. Cooling Cycle, N. Sifter-bagger.

- Low building height
- Forced removal of the powder from the chamber (i.e., controlled residence time)
- Easy access for cleaning
- Hammers or vibrators on the chamber are not required.

(b) Vertical Drying Chamber

In this type of dryer, the greater portion of the powder is removed by gravity.

The advantages of a conical bottom chamber are :

- The incorporation of two or three stage drying.
- A requirement for the separation of the powder into a coarse fraction from the chamber and a fine fraction from the cyclone. By recycling the fines to the atomisation zone, a dust-free party agglomerated powder is obtained.
- Powder with a high fat content (exceeding 35-40%) has tendency to stick to the interior surface of the ducts and cyclones, and therefore, the greater portion of the powder must be discharged directly, from the chamber, so that only the small quantity contained in the air will pass through the ducts and cyclones.

Vertical chambers can be conical-based, inverted-cone type (w-shaped), flat-based usually with sweeper or conical-based nozzle tower type (Tall from dryer) drying chambers, which are designed to discharge all the product with the exhaust air. They are, sometimes, provided with a rotating air-broom. The air broom consists of a bent tube which slowly rotates around the axis of the chamber. The air broom is so shaped that it moves along the chamber wall and at a short distance from this. The tube is, in its whole length, provided with opening. An air current is blown at high velocity against the wall, removing the powder settled there.

Usually, chambers are equipped with electric hammers on the outside. These hammers/vibrators prevent powder from settling on the wall inside.

ATOMIZER

Milk atomizers serve a basic function in spray drying. They provide a high surface-to-mass ratio; thus enable quick heat transfer with a high evaporation rate. One litre of milk in a spherical shape has a surface area of about 0.05 m^2. If this quantity of milk is automized in the spray tower, each of the small droplets will have a surface area of 0.5-0.15 mm^2, i.e., atomizing increases the specific area by a factor of about 700.

Atomization parameters influence some important properties of the final product. These may include texture grain size, bulk density, solubility, wettability and density.

In commercial spray drying, three types of atomizers are employed; centrifugal or rotary atomizer, pressure nozzle and prentice or two fluid nozzles (Fig. 5.11 & Fig. 5.12).

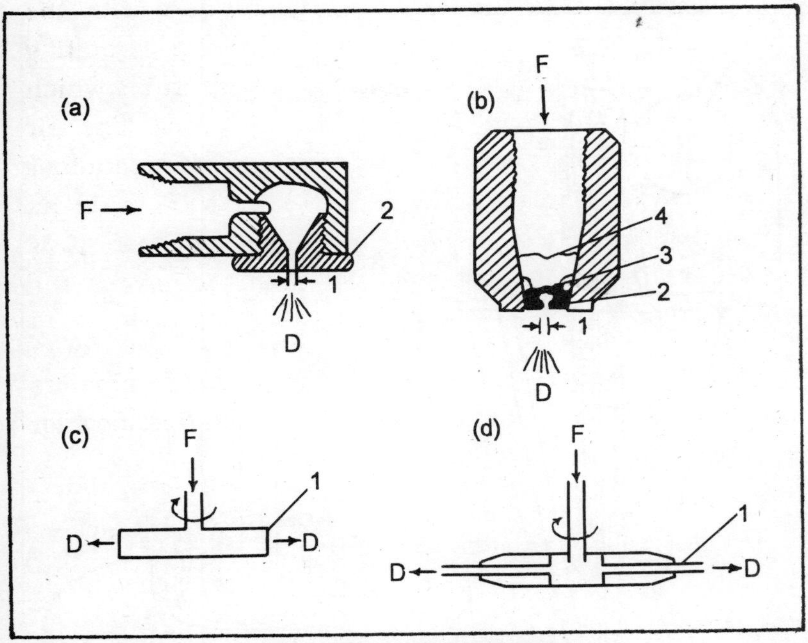

Fig. 5.11 Atomizer designs

(a, b) Pressure nozzle : 1. Diameter of orifice, 2. Orifice plug, 3. Groove, 4. Core; (c, d) Centrifugal atomizer : 1. Vertical slots or spokes, F. Feed, D. droplets.

Pressure Nozzle

The pressure at the nozzle determines the particles size. At high pressure, upto 30 MPa (300 bar), the powder will be very fine and have a high density. At low pressure 5-20 MPa (50-200 bar), large particles will be formed and the fine content will be lower. The pressure is built up by means of multi-plunger high pressure pumps. Pressure atomizers were developed and used in North America.

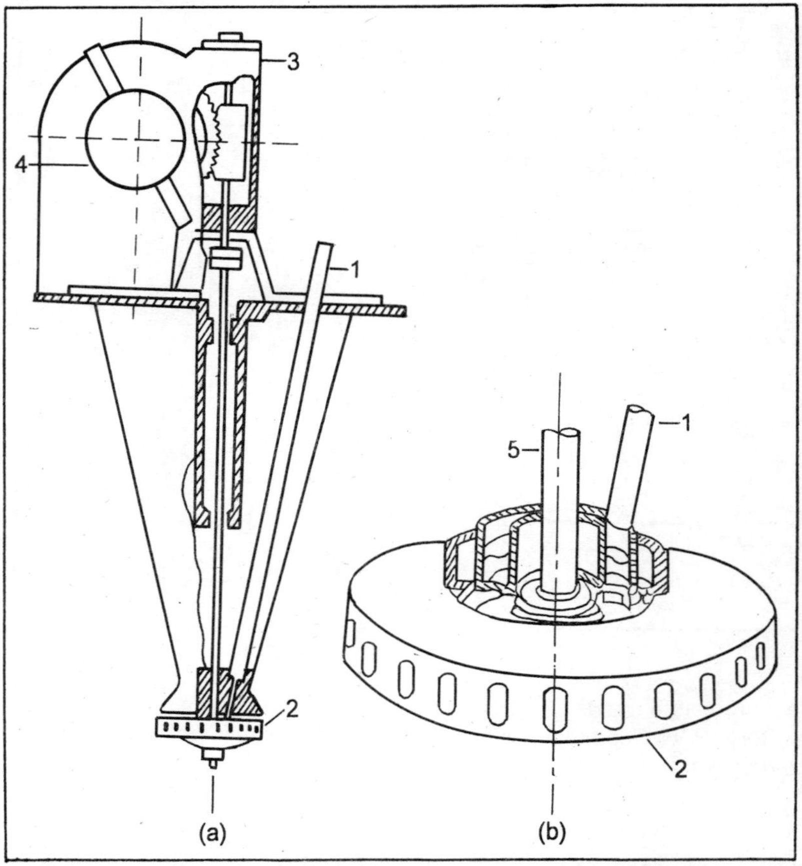

Fig. 5.12 : Schematic diagram of disc type centrifugal atomizer

(a) Diagram showing arrangement of a centrifugal disc atomizer.

(b) Atomizing disc.

1. Milk inlet pipe, 2. Atomizer wheel, 3. Housing for worm wheel mechanism, 4. Driving pulley to motor, 5. Shaft carrying atomizer wheel.

Centrifugal Atomizers

These are used extensively. They include spinning disc, rotary cup and spinning nozzle of which spinning disc is most popular in spray drying of milk. Centrifugal atomizers originated in Europe. Where they are predominantly used. Centrifugal atomizers in advantageous is drying viscous material and suspensions.

In principal disc type centrifugal atomizer consists of an atomizer wheel, rotating at high speed 10,000-20,000 rpm. on a shaft by a vertical motor (Fig. 5.12). One of two feed tubes take the feed milk down to a ring shaped liquid distributor from where the milk runs down into the wheel or disc. Centrifugal atomizer can be used successfully with viscous liquids as chances of blockages of the disc opening are rare.

Pneumatic Atomizer

In pneumatic atomization (Fig. 5.13), the energy for atomization is provided by the compressed air. The atomization is not uniform as other atomizers and

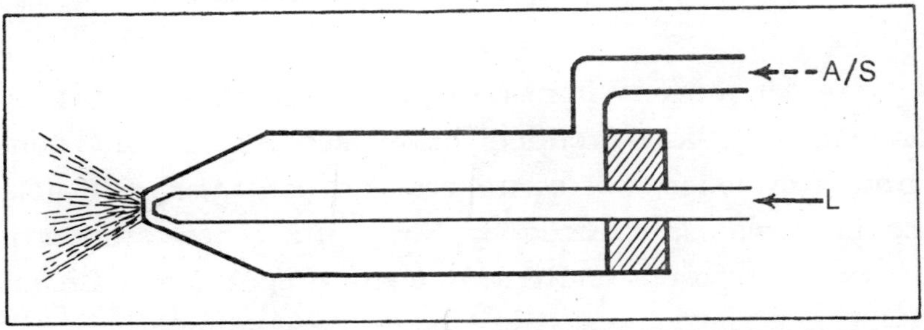

Fig. 5.13 Pneumatic atomizer A/S = Airosteam. L = Liquid.

consumes highest amount of energy. It is not used in commercial milk dryers. They often applied in small diameter pilot plant chamber where both fine and medium coarse atomization can be handled.

POWER CONVEYING SYSTEM

Pneumatic conveying is a common technique for transporting milk powder from the base of the drying chamber and powder separators to a packing or storage area. There are two systems of pneumatic conveying.

(i) Low pressure or vacuum system : In this system, the centrifugal form is mounted after the transport cyclone. Transport velocities of 20-25 m/sec. are used. The air-product weight ratio is high. Usually large duct diameters are used.

(ii) High pressure system : The high pressure system uses a rotary type blower placed before the powder inlet and employs high air velocities.

POWDER RECOVERY SYSTEM

Powder recovery from the spray dryer exhaust systems for food and dairy products have been traditionally done through cyclones, wet scrubbers and bag filter. 'CIP able' bag filter was introduced by Niro setting new standards for powder recovery and environmental regulations. (Golani, 2005)

Cyclones

The principle of cyclone separation is based on centrifugal force. The air enters the cyclone with high velocity through the entering duct, the diameter of which is approximately four times smaller than the diameter of the cyclone cylinder. The air velocity decreases allowing powder to fall by gravity to the bottom of the cyclone, where it is continuously removed. Higher air velocity induces higher pressure drop, thus enabling separation of the small powder particles. In application, there is a system of several cyclones with large diameter and one cyclone with small diameter at the end, which serves to separate fines.

The air velocities at inlet upto 30 m/sec. can be used. Pressure drop in cyclones ranges from 5 to 15 mm kg.

Bag Filters

They are, sometimes, employed as secondary collectors after cyclones. Bag filters are of cotton, wool or plastic. Cotton and wool bag filters should be used at temperatures upto 90°C. Other bag filters may withstand temperatures of 120°C to 130°C. Bag filters are installed in multiples and are mechanically shaken on a pre-determined time cycle. With the correct fabric weave/ micron particle size can be collected. Collecting efficiency of 99 per cent is obtained with 1-2 micron sizes. Air velocities used are low 0.5–7.5 m/min. Pressure drop ranges from 3 to 15 mm Hg.

Wet Scrubbers

The use of a wet collector or wet scrubber after the cyclones to trap the greater portion of the solids by scrubbing the air with water. The water is recirculated and periodically renewed.

The Disadvantages

- Milk solids are not recovered
- Disposal of the scrubbing water
- Scrubbers are ideal for rapid growth of micro-organisms.

To overcome the problem of recovery of solids and disposal of scrubbing medium, the milk scrubber is introduced. Unconcentrated milk is used as a scrubbing medium instead of water to absorb the solids from the air. The milk goes from the scrubber to the evaporator, and the solids recovered will be contained in the concentrate to the dryer and will, thus, increase the yield. However, microbial contamination problem was not solved.

SANI CIP Bag Filter

The trends are changing now, where the environmental and sanitary issues have been in focus, the legislation and market would demand stringent recovery systems. After 10 years of R & D, 'CIPable' bag filter was introduced by Niro A/s Denmark, setting new standards for powder recovery and environmental regulations.

The Niro SANI CIP bag filter, consists of cylindrical bag housing with spiral shaped air inlet and comprises broadly three sections :

- Clean air plenum on the top with reverse jet nozzles
- Bag filter section with exhaust air inlet from chamber and internal/external CIP nozzles.
- Conical bottom for fluidized product discharge.

The filter bags are made from a special 3 layer gradient polypropylene material, which is heat-treated to give a special dust releasing surface Niro A/S has developed a special design of reverse jet air nozzles for SANI CIP bag filter.

During operation, the product collected on the outside of the filter material (bag) is removed by a compressed air jet stream blown into each bag through a special reverse jet air nozzle positioned above each bag. A jet is formed which draws air from the clean air plenum into the bag as well, thereby saving compressed air. This is an efficient and sanitary solution. The bags are clean blown individually or in multiples together resulting in an even discharge of powder and using higher air to cloth ratios.

The CIP of the clean air plenum is done through a spray ball. The reverse jet nozzle has, furthermore a dual purpose during CIP and cleans the bag from inside towards the dirty side (outside) as a part of internal bag CIP.

Clean water is injected inside of the bag through a reverse jet nozzle and water is atomized by compressed air. Powder that has penetrated into the bag material is forced out towards the dirty side by the water spray.

It is very difficult to clean entrained powder from outside only. The hole plate CIP cleans the bottom side of the hole plate and the snap ring area of the bag using a specially designed nozzle also with dual purpose. The CIP is followed by bag drying.

Advantages of Niro 'SANI CIP Bag Filter

1. Low pressure loss across the bag filter and, thus, the entire exhaust system, i.e., reduced energy consumption.

2. Minimal stock losses and higher yield of raw materials due to second grade products.

3. Designed for optimum air to cloth ratio and powder load.

4. Compliance to sanitary standards with complete CIP of the entire bag filter by means of :

- Internal bag CIP from inside towards the outside (dirty side of the bag).
- CIP of clean air plenum
- CIP of hole plate
- CIP of housing
- Reduced dry-out time after CIP.

POWDER COOLING SYSTEM

The dried powder should be removed and cooled immediately after the completion of drying to minimize the heat damages, to prevent the powder from lumping and to enhance the keeping quality. The temperatures of powder after the completion of drying is about 55°C to – 60°C. The best methods of cooling are those which cool the powder during its removal from the dryer. Employing conditioned air for preumatic conveying provides necessary cooling effect. Powder is cooled to temperature of 30°C or lower.

EFFICIENCY OF DRYING

Efficiency of drying can be calculated by using the equation :

$$\text{Efficiency of Drying } (\eta) = \frac{T_{in} - T_{out}}{T_{in} - T_{amb}}$$

where, T_{in} is the temperature of the inlet air into the dryer, T_{out} is the temperature of the outlet air from the dryer and T_{amb}, is the ambient temperature of the air.

DIFFERENT TYPES OF SPRAY DRYING

Single Stage Drying

The simplest type for producing a powder consists of a drying chamber with an atomization system, the air

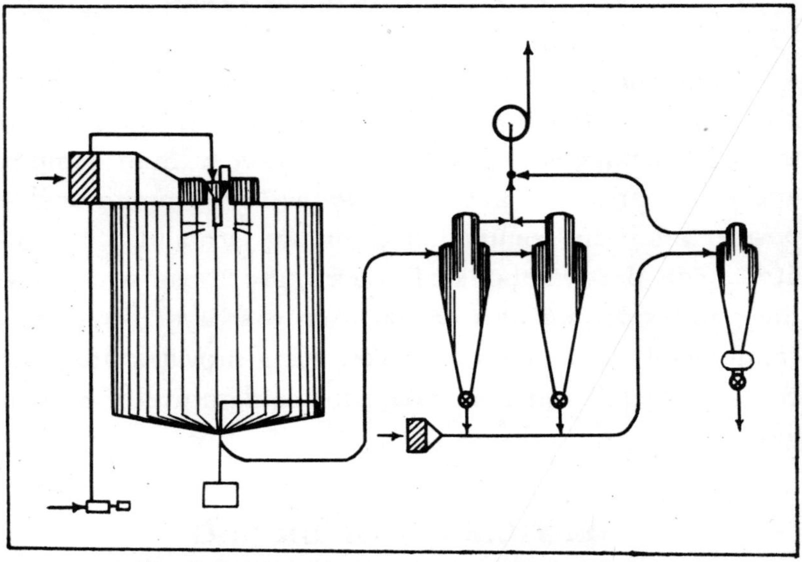

Fig. 5.14 Single stage spray dryer; drying chamber with flat bottom

(***Courtesy*** : APV Anhydro A/S Denmark)

heater, a system for collecting the finished powder from the dry air and a fan which sucks the necessary amount of air through the entire system. In single stage dryer (Fig. 5.14). The entire drying process takes place in a single unit, the drying chamber. A powder with a small grain size and a high fine content is produced here.

Two-Stage Drying

In two-stage dryer, the product is dried upto 7-15 per cent moisture in one step and the remainder of the water is evaporated in the second step. In the first stage of drying, the product is dried at a lower outlet temperature and a higher moisture content. In the second stage, the high moisture product is transferred into a fluidized bed drier for final drying. The product is treated very gently in the second stage of drying to obtain the desired final moisture content which, in turn, controls the powder's physical and chemical properties. High evaporation rates are obtained with gentle heat treatment and a large temperature differential between the inlet and outlet air is maintained. Temperature of the product after the first stage of drying is slightly higher than that of the powder obtained by single-stage drying. Higher product temperature is the direct result of the high temperature differential but the final quality of the powder is not lowered. A typical configuration for a two stage dryer is shown in Fig. 5.15. Higher evaporation rates due to the larger temperature differential result in much lower energy consumption than Single stage drying.

Two stage drying of milk and dairy products offers the distinct advantages of producing a superior product at lower energy levels. More specifically, two stage drying processes have the following advantages over single stage dryers.

- Efficient evaporation is achieved
- Higher production rates are possible
- Energy costs are reduced substantially
- Thermolabile substances can be dried with minimal effects on functional properties.
- Large particle sizes and agglomerates can be dried without distorting or crushing.

Fig. 5.15 Two-stage dryer with solids recovery and heat recovery.

1. Textile filter, 2. Heat exchanger for heat recovery, 3. Booster in flue gas stack, 4. airpreheater, 5. fluid lift system to powder silo.

(***Courtesy*** : APV Anhydro A/S Denmark)

- During the second stage of drying surfactants such as lecithen can be applied to improve rehydration properties.
- Use of fluidized bed second-stage provides easy and accurate operational control.
- The system is flexible and has a multitude of applications.

Comparison of one stage and two stage drying system is given in Table 5.2.

Table 5.2

Comparison of One Stage and Two Stage Drying System

Drying System	One Stage Inlet temp. 200°C	Two Stage Inlet temp. 200°C	Inlet temperature 230°C
1	2	3	4
Spray dryer (Ist stage)			
Evaporation in chamber, kg/h	1150	1400	1720
Powder from chamber			
6.0% moisture kg/h	—	1460	1790
3.5% moisture kg/h	1140	—	—
Energy Consumption			
Spray drying total M Cal	1818	1823	2120
Energy/kg. powder, Kcal	1595	1250	1184
Fluid Bed (2nd Stage)			
Drying air, kg/h	—	3430	4290

1	2	3	4
Inlet air temperature, °C	—	100	100
Evaporation influid bed, kg/h	—	40	45
Powder from fluid bed			
3.5% moisture, kg/h	—	1420	1745
Energy Consumption, kW	—	20	22
Energy Consumption,			
total in fluid bed Mcal	—	95	115
Total Plant			
Energy Consumption, Total Mcal	1818	1918	2235
Energy/kg. powder total Kcal	1595	1350	1280
Energy relation	100	85	80

Basis : Same drying Chamber size with inlet air flow = 31500 kg/h.

Product : Skim milk, 48% solids in concentrate

Source : Evaporation, Membrane Filtration, Spray Drying – North European Dairy J. (1985). Copenhagen, Denmark.

Three-Stage Drying

Three-stage drying is an extension of the two-stage concept, developed to achieve greater savings in process costs and to meet various product quality demands more effectively.

In three-stage drying, the second stage is a fluid bed placed at the outlet of the conical part of the drying chamber. this fluid-bed is static (non-vibrating), and the semi-dry powder drops over an adjustable weir into an

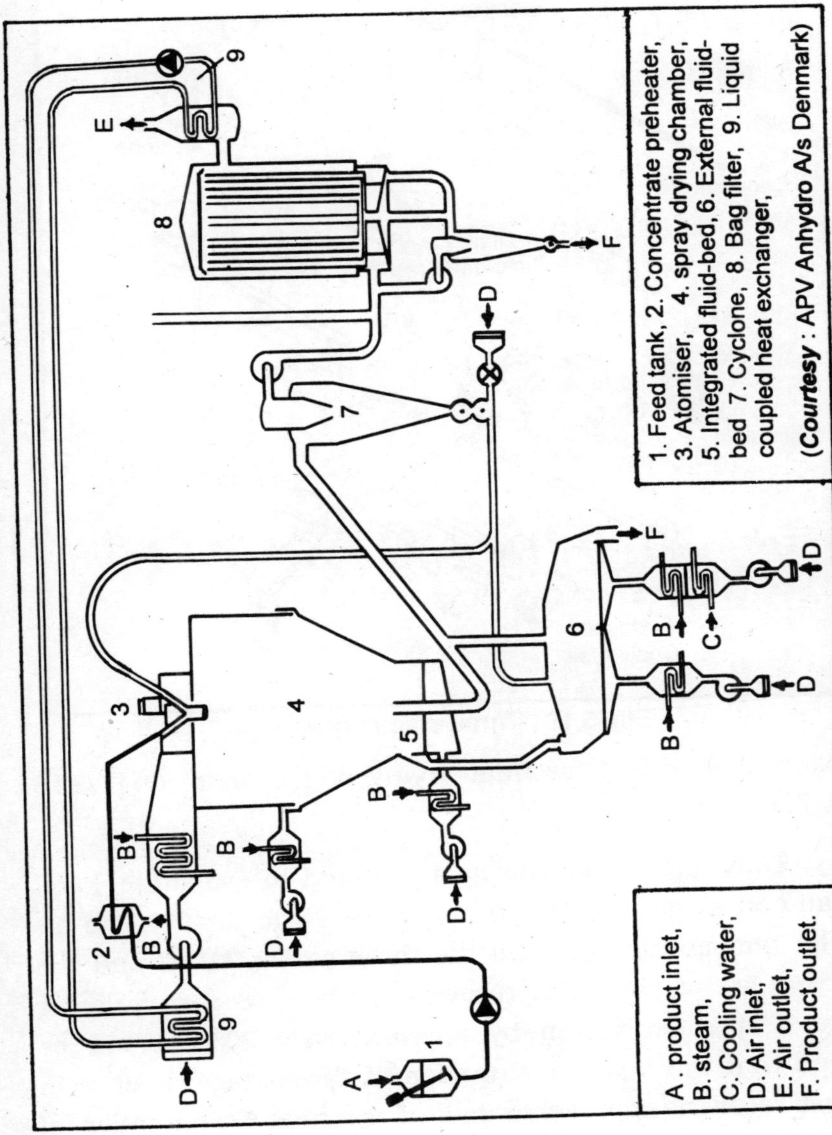

Fig. 5.16 Three-stage dryer with solids recovery and heat recovery

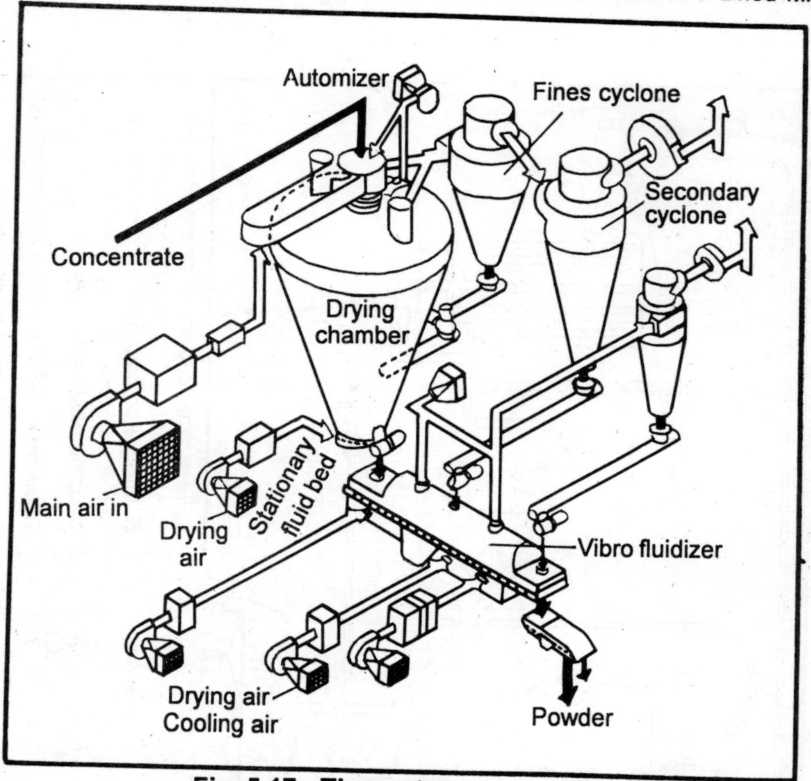

Fig. 5.17 : Three stage spray dryer

external fluid-bed for final drying and cooling. (Fig. 5.16 & Fig. 5.17)

Table 5.3 shows the heat required to evaporate 1 kg. water in single, two-stage and three-stage dryers. The air inlet temperature is in all the three cases, 215°C and the dryers are without heat recovery. As heat recovery reduces the heat consumption by approximately 20%, it will be seen that the heat consumption of three-stage dryer with heat recovery approaches half of the heat consumption of a single-stage dryer without heat recovery. This clearly illustrates the progress made in energy conservation.

Table 5.3

Heat Required to Evaporate 1 kg Water.
Air Inlet Temperature 215°C.

Types of dryer	Kcal
Single-stage	1200
Two-stage	980
Three-stage	850

The advantages of three-stage drying

- Reduced heat consumption
- Reduced space demand
- Improved powder quality
- The amount of air required to obtain a certain capacity is the smallest possibility when three-stage drying is used and thereby the cost of a textile filter for cleaning the outgoing air is reduced.

DIFFERENT TYPES OF DRYERS

Tall Form Dryers

The tall form dryer (Fig. 5.18) has pressure nozzle atomisation by means of one or more verticle nozzles. Vertical nozzles requires a tall drying chamber of small diameter; the height is approximately five times the diameter and this chamber configuration does not permit the use of a rotating atomiser.

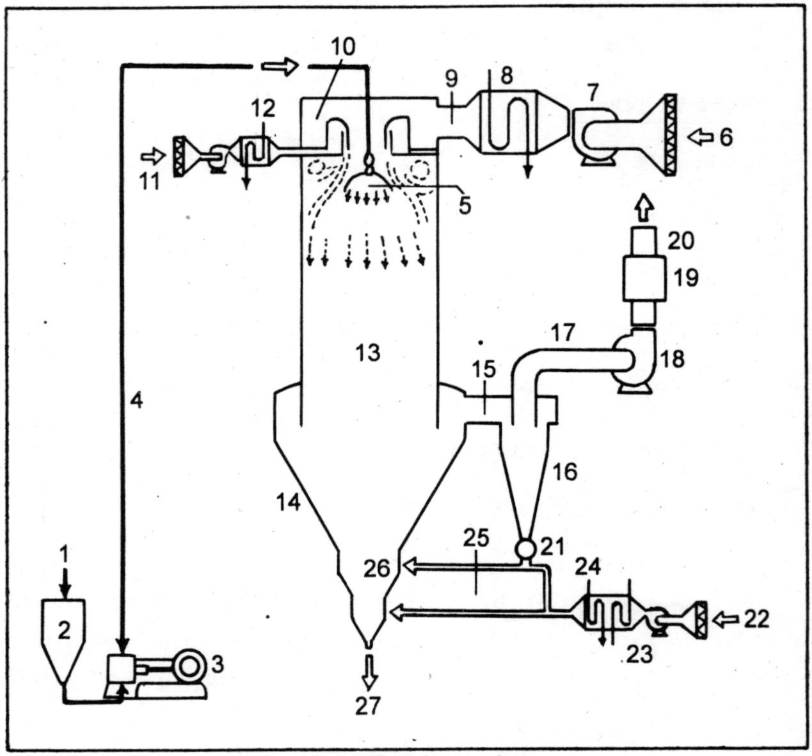

Fig. 5.18 Tall form dryer

1. Concentrate, 2. Feed vat, 3. High pressure pump, 4. Feed pipes, 5. Pressure nozzle, 6. Main air inlet, 7. Fan, 8. Air heater, 9. Hot air duct, 10. Air distributor, 11. Secondary air inlet, 12. Air heater, 13. Drying chamber, 14. Separating cone, 15. Outgoing air duct, 16. Cyclone collector, 17. Duct, 18. Exhaust fan, 19. Silencer, 20. Exhaust stack, 21. Rotary valve, 22. Cooling air inlet, 23. Air cooler (dehumidifier), 24. Air heater, 25. Cooling air ducts, 26. Powder cooling zone, 27. Powder discharge.

(***Courtesy*** : Morinaga, Japan)

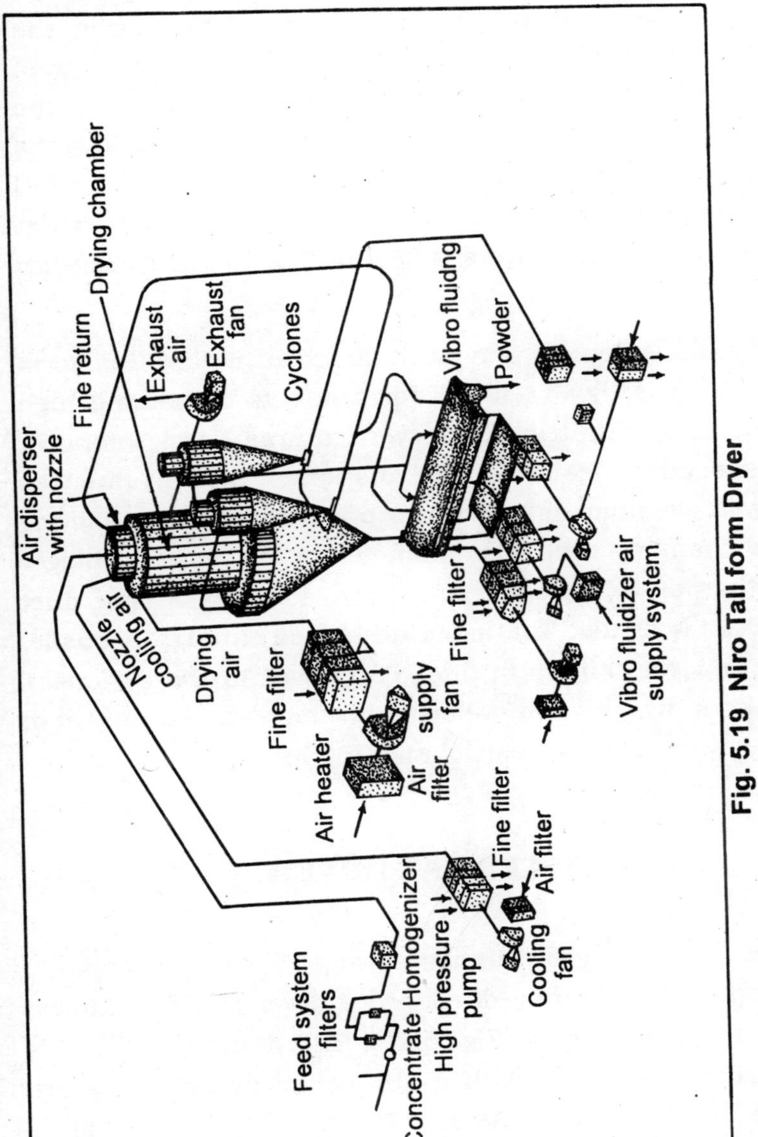

Fig. 5.19 Niro Tall form Dryer

Courtesy : NIRO A/S Denmark. GEA processing Engineering (India) Ltd.

The absence of a swirl movement of the air in the chamber is claimed to result in a true co-current flow of air and product. This gives a short residence time, the advantage of which is somewhat counteracted by the higher outlet temperature required to achieve the desired moisture content of the powder. The elevated outlet temperature has an adverse effect on the thermal efficiency of the dryer.

Fig. 5.19 depicts a typical sketch of a Niro Tall form dryer which is ideally suited for products containing high fat and carbohydrates. This dryer features Nozzle atomizer and specially designed Air disperser which ensures downward laminar air stream to prevent particles adhering to the chamber wall. The diameter of the cone is bigger than that of cylindrical part, thus, forming a ring duct termed as a bustle. The air is exhausted through a bustle, which ensures minimum fines carry-over to the cyclones/ bag filters by the low velocity of the exhaust air and, thus, low product loss (Golani, 2005).

FILTERMAT DRYER

The University of Minnesota and Pillsbury Company, Minneapolis have jointly developed a dryer which is known as the Filtermat Dryer (Fig. 5.20). The drying chamber is rectangular (box type) with a number of verticle pressure nozzles in the ceiling. As the height of the chamber is insufficient for drying to the final moisture content, the moist product builds up into a mat on a porous moving belt. The drying air goes through this layer of powder which acts as a filter for the exhaust air. The moist

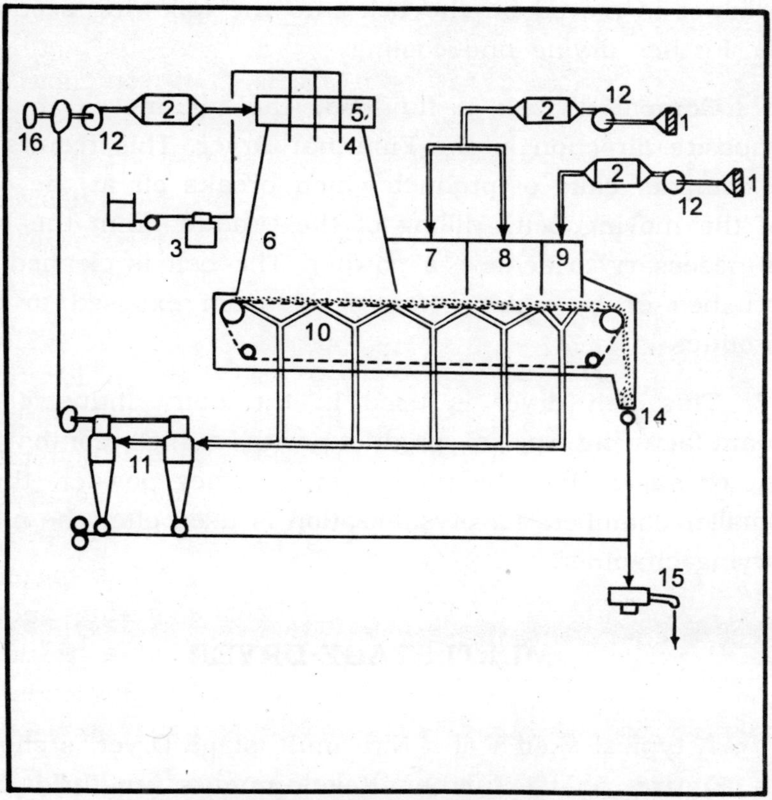

Fig. 5.20 Integrated filtermat belt drying unit

1. Air filters
2. Heater/scooler
3. High pressure pump
4. Nozzle arrangement
5. Air distributor
6. Primary drying chamber
7. Retention chamber
8. Final drying chamber
9. Cooling chamber
10. Belt assembly
11. Cyclone arrangement
12. Fans
13. Fines recovery system
14. Powder discharge
15. Sifting system
16. Heat recovery system

product is moved on the belt into the following sections for further drying and cooling.

Compared with a fluidbed, the air moves in the opposite direction in the Filtermat Dryer. This results in a compact cake of product which breaks off at the end of the moving belt. Milling of the product from the belt is necessary to achieve a powder. The belt is cleaned by brushes or by water before it is again exposed to the product.

This belt dryer is used in the dairy Industry for manufacturing special products with a high content of fat or sugar. For the production of whey powder, three smaller chambers for crystallisation is used after the main drying chamber.

MULTI-STAGE DRYER

A typical sketch of a Niro multi-stage Dryer is shown in Figure 5.21. Further developments in fluid bed techniques have resulted in a static fluid bed designed as circular integrated part of drying chamber. Three stage drying combine all the advantages of extended two stage drying using spray drying as the primary stage, fluid bed drying on a static fluid bed as second stage and external fluid bed as the third. The critical phase of a two stage drying occurs when the wet particles come in contact with the surface of the equipment. Depending upon the composition and sticking temperature, the powder may start to stick to warm metal surface and forms deposits and lumps.

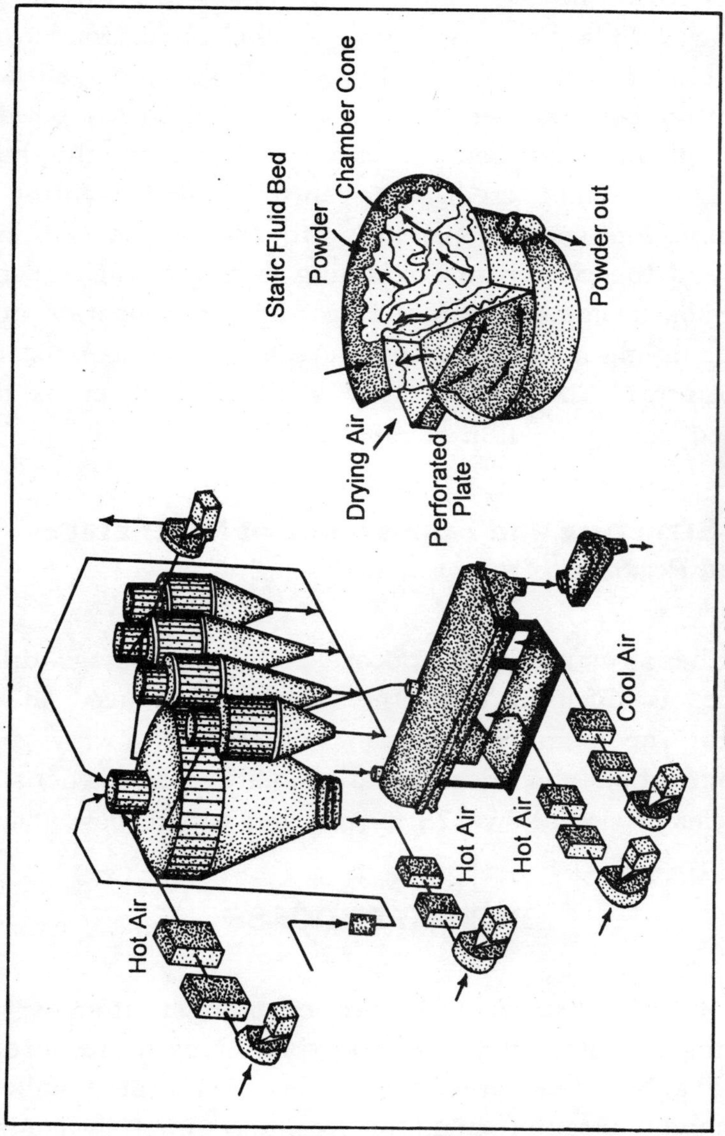

Fig. 5.21 Niro multi-stage dryer

(***Courtesy*** : NIRO A/s Denmark) GEA Process Engineering (India) Limited

The basic idea behind this dryer design is to operate the first drying stage with high moisture levels (as compared to a two stage) and at the same time avoid any contact of the wet powder with the chamber surface by introducing powder directly into the fluidized powder layer of the fluid bed. In multi-stage dryer, the high velocity air inlet creates a venture effect creating a vacuum. Fine particles along with the air sucked into inflow of the concentrate resulting in a very high degree of agglomeration. The multi-stage dryer can operate at a higher drying air inlet temperature but a low exhaust air temperature. This makes it an economical dryer as the applied energy is utilized efficiently.

The Structure and Appearance of Multi-stage Dried Powder

The structure and appearance of multi-stage dried powder is different from the powders produced in all dryers. The main characteristics related to a very good agglomeration and mechanical stability, low fractions of particles of size below 125 micron and superb flow ability.

DRYING PROCESS

Milk is dispersed in the centrifugal atomizer at rotating speed of 10,000-20,000 rpm. or by a pressure of 17-25 MPa in the pressure nozzles. Evaporated milk is discharged through small orifices and dried in contact with hot air. In this way, fine particles, uniform in size, having a diameter of 20-150 mm (most of which are in

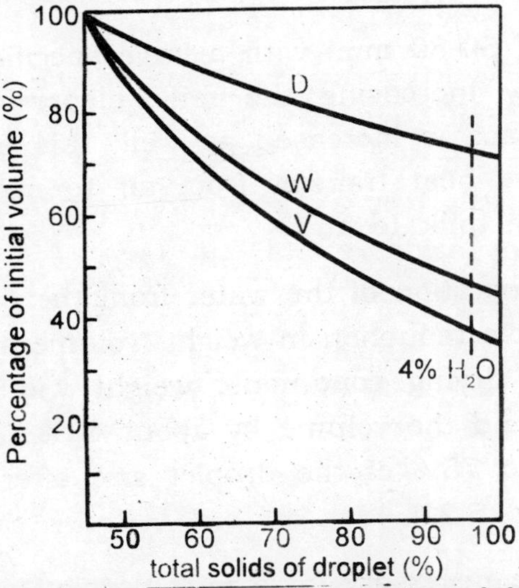

Fig. 5.22 : Decrease of dispersed droplets weight (H), volume (V) and diameter (D), during drying of skim milk wtih 45% solids under ideal drying conditions

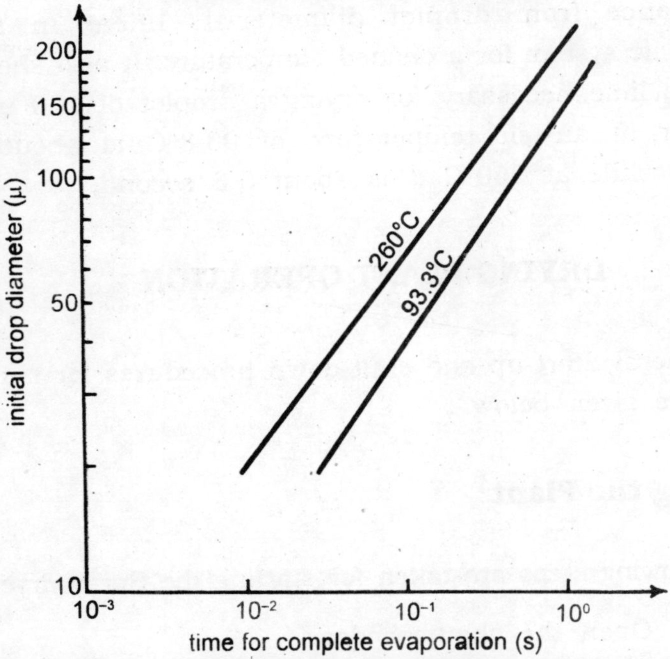

Fig. 5.23 : Time of evaporation of water drops in spray dryer

the range of 50-80 mm) with a large specific surface are obtained. By increasing the milk dispersion rate, the specific surface is increased as well; this enables rapid and intensive heat transfer from air to milk and mass transfer from milk to air.

The evaporation of the water from the droplets leads to considerable reduction in weight, volume and diameter. Under ideal drying conditions, weight will decrease by about 50% and the volume by about 40%. The diameter is reduced to 75% of the droplet size after leaving the atomizer (Fig. 5.22).

The time necessary for a milk droplet to be dried in a spray dryer depends mainly on droplet diameter and air temperature. Fig. 5.23 shows the drying time dependence from droplet diameter is linear in the logarithmic system for a defined temperature. It also shows that the lime necessary for drying a droplet of 100 mm diameter at an air temperature of 93.8°C is about 1 second, while at 260°C it is about 0.3 second.

DRYING PLANT OPERATION

General Start-up and shut-down procedures for spray dryer are given below :

Starting the Plant

Following steps are taken for starting the Spray dryer :

- Open the chamber door

- Start the exhaust fan and cooling air fan (for the cooling air disperser)
- Start the supply fan
- Close the chamber door and air sweep doors
- Start the rotary values, vibrating device and air conveyor fan
- Start the atomizer
- Admit steam
- When the outlet air temperature is raised to 85°C-100°C the feed pump is started and water is fed to the atomizer. Adjust the outlet air temperatures at around 95°-100°C adjusting the feed of water.
- When the inlet air temperature obtains to 190°C-200°C and other conditions are stabilized, switch over to concentrate.
- Adjust outlet temperature to 90°C-95°C.
- When the product is fed to the plant, the outlet air temperature should be continuously observed. The moisture in powder is adjusted as required. The normal flow of powder from the plant should be under constant observation.

Shut-down

The following operations are carried out when stopping the plant.

- Change over to water from product. Feeding of water should be controlled (decreased) so as to maintain the outlet air temperature, and to avoid soiling the chamber.
- Close the steam value of the air heater.
- When the inlet air temperature comes down to about 120°C-150°C stop the feed pump.
- Open the drain cock provided on feed line to the atomizer.
- Stop the atomizer.
- Open the door and hook it to prevent complete closure of the door.
- Stop the air supply and cooling fans.
- Stop chilled water to the powder cooling system.
- Allow air sweeping of the chamber.
- When all the powder is carried out of the plant the exhaust fan, conveyer fan, rotary values, etc. are stopped.
- Close the chamber door and dampeners on the air ducts.

It is important that when the exhaust fan is running without air supply, the chamber door should be kept open. Otherwise the reduced pressure in the chamber may seriously damage the chamber lining.

Problems in Spray Dryer Operation and Remedial Measures

Problems & their Causes	Remedial measures
(A) Incorrect flow rate	
1. Incorrect feed concentration	Check up feed concentration
2. Incorrect atomizer pressure or speed.	Check up and take necessary action.
3. Plugged liquid passages, especially in pressure nozzles	Clean the passages, use filter before atomizer.
4. Incorrect atomizer selected with respect to its rating.	Check up and use correct atomizer.
(B) Improper Spray Pattern	
• Spraying very high concentrated milk, especially in the case of pressure nozzle.	Decrease milk concentration to acceptable level.
• Atomizer plugged with milk solids or extraneous materials.	Clean the atomizer and reassemble properly.
• Atomizer incorrectly assembled.	Reassemble properly.
• Atomizer damaged	Replace atomizer.
• Low pressure or Speed	Adjust pressure or speed.
• Very high pressure or Speed	Adjust pressure or speed.
(C) Coarse powder particles	
• Feed too concentrated	Adjust concentration or heat the feed.
• Lower pressure or slow speed.	Increase pressure or speed.
(D) Very fine powder particles	
• Feed too dilute	Increase concentration.
• High pressure or speed	Decrease pressure or speed.
(E) Rapid flow increase	
• Atomizer erosion	Check up and take necessary action.
• Leakage in the system.	Check up for leakage in atomizer and auxiliary plumbing.
(F) Sudden drop inflow rate	
• Plugging of atomizer	Clean atomizer, use strainer
• Blocking of feed line	Clean feed line.
• Manufacturing of pumps and motors.	Check up and take corrective steps.

Problems & their Causes	Remedial measures
(G) Overheating of pumps and motors	
• Pumps or motors operating inefficiently	Check up and take necessary action.
• Pumps or motors underdesigned.	Use correct size of pumps or motors.
(H) Plugging of pressure nozzle orifice.	
• Contaminants in milk	Use strainers.
• Too high concentration of milk.	Use milk of correct concentration.
• Low milk velocity	Use high pressure
• Too small nozzle orifice	Use nozzle of larger orifice, if possible.
(I) Erosion of pressure nozzle	
• To high feed concentration	Check up and follow-up.
• To high atomization pressure	Reduce pressure
• Feed containing foreign materials	Use strainers before atomizer
• Cavitation due to high-feed temperature and air leakages.	Limit the feed temperature and check up for air leakage
(J) Corrosion of Atomizer	
• High feed temperature	Lower feed temperature
• Improper detergent used in washing solution	Use proper detergent in washing solution at proper temperature.
(K) Deposition on atomizer or dryer wall	
• Incorrect spray pattern.	Adjust pressure or speed.
• Coarse droplets	Adjust concentration or heat the feed.
• Air disperser vanes wrongly set	Adjust the air disperser vanes until the humid deposits disappear.
(L) Incomplete drying	
• High milk flow rate	Decrease the flow rate.
• Coarse droplets	Adjust concentration of heat the feed.
• Low air outlet temperature	Adjust air outlet temperature
• High air inlet humidity	See Problem R
• Incorrect spray pattern	See Problem B
• Leakage	Cehck up for leakage and take remedial step.

Problems & their Causes	Remedial measures
(M) Low moisture in powder	
• Low flow rate	Adjust flow rate
• High air outlet temperature	adjust
• Fine atomization	see problem D
(N) Burnt powder particle	
• Air temperature too high	Adjust temperature of air.
• Very fine atomization	See Problem D.
• Improper and delayed cooling of powder	Cooling should be done during removal of powder and below 30ºC.
(O) Sooty Powder	
• Insufficeint air filteration	Clean Or change the filter.
• In direct fired heating units (DFHU), misfiring	Check up and follow.
• In DFHU, incomplete combustion	Ensure complete combustion.
• In DFHU, failure of the gas fan.	Check up and rectify the mistake.
(P) Lumpiness in powder	
• Excessive moisture in powder	See. problem L.
• Insufficient and slow powder cooling.	Ensure fast cooling of powder to 30ºC or lower.
• Failure of cooling arrangement.	Check up
(Q) Dusty air exhaust	
• Inefficient powder separation	Check up and follow.
• Too fine atomization	See problem D.
• Low powder moisture	See problem M.
• Increased load.	Adjust
(R) High Air humidity	
• This problem arises usually during humid summer weather (June to Sept.)	Milk feed rate may be decreased, air outlet temperature increased or air inlet temperature increased, if permissible.
(S) Reduced evaporation capacity	
Reduced flow of air.	Adjust the fan speeds, check motors, dampeners etc.
Too low air inlet temperature.	Check the air heater and the burner.
Leakage in the plant.	Inspect plant and repair damages.

CLEANING OF SPRAY DRYING PLANT

Normally, after every operation air sweeping is done to remove the powder sticking to the walls of the chamber, air filters, rotary valves, exhaust fan and conveyor fan impellers. However, it will become necessary to clean the system properly after a few runs. Two types of cleaning systems are employed :

Dry Cleaning

When all the powder is not removed by air sweeping, it is necessary to remove the powder sticking to the system by manual brushing. In this, the chamber wall, cyclones and conveyor ducts are swept manually by the help of a brush and cleaned.

CIP Cleaning

Increase in labour and material costs and the damage to equipments during dismantling and assembling have increased interest in in-place cleaning. The dryers are getting bigger and more complex, automatic cleaning in the shortest possible time is required.

CIP includes pre-rinse, caustic wash, middle rinse, acid wash and final rinse and finally dryout. For sterilising purposes hot air is applied to the wall of the dryer for 10 min. at 90°C.

CIP is done by filling spray balls (Fig. 5.24) in the plant in such a way that the spray balls cover the entire

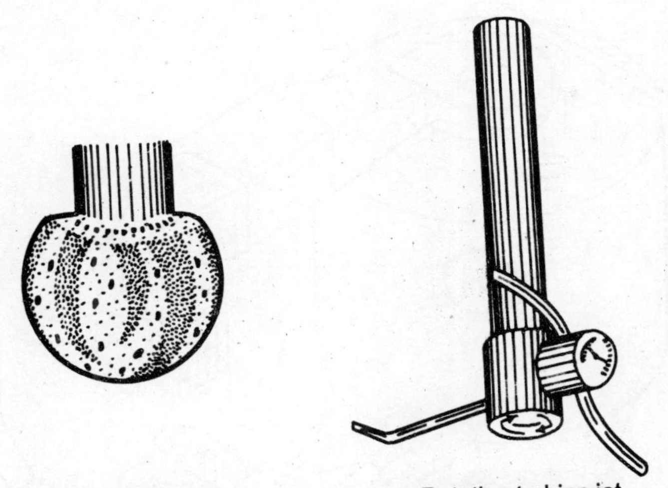

Fig. 5.24 Spray cleaning devices

interior surface. They are permanently placed in ducts, cyclones and fluid-beds (above and below the perforated plate) (Fig. 5.25). The spray balls are stationery and in the large dryer 60-80 spray balls are required. It is possible to leave the spray balls in powder ducts, cyclones and fluid beds because when the plant is in operation, heated, purging air is blown through the entire CIP-system, thereby preventing the product from penetrating into the spray balls.

Only the drying chamber is not cleaned by a built-in-cleaning device. The chamber is cleaned by removing the atomizer and lowering a rotating jet tank cleaner (Fig.

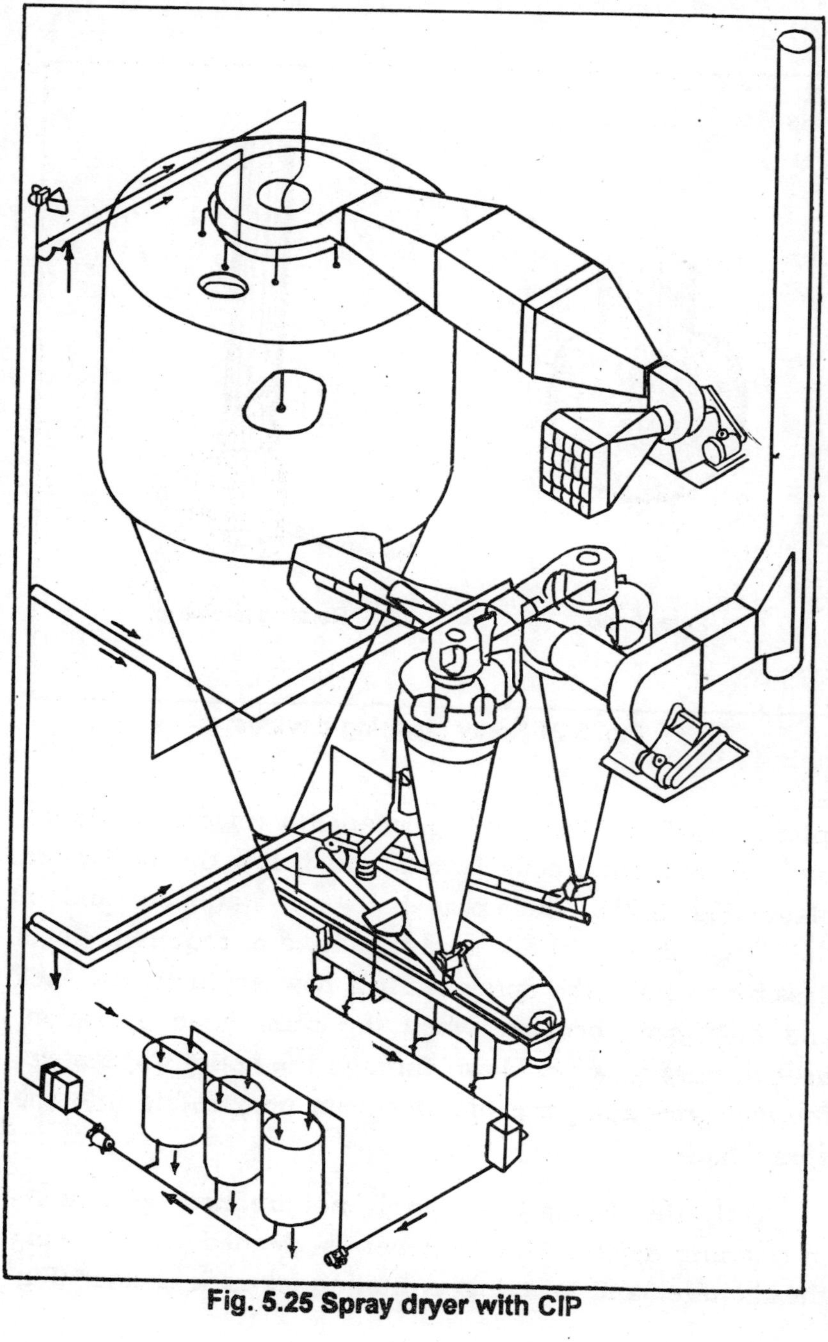

Fig. 5.25 Spray dryer with CIP

5.24) down into the chamber. The rotating-jet cleaner is very effective, as the water hits the surface at high pressure, thereby producing a mechanical action for the removal of wall deposits.

Cleaning water from first rinse contains some milk solids. It may be used for animal food. Cleaning solution is normally connected in a tank for re-use.

Fire and Explosion : Possibilities and Prevention in Spray Drying Plant

Safety does not mean absence of danger but a device to tackle it effectively. Drying involves high temperature processing and at 260°C anything can happen in terms of explosion in the dryer or heavy product loss. Thus, efficient control of dryer is a must, so that no ignition and resultant mishap occur in a drying plant. The hazard analysis at critical control point must be pondered upon beforehand to prevent any untoward incident in a drying plant.

Possibilities of Ignition and Explosion

1. **Self-ignition or Spontaneous combustion** : Self-ignition or spontaneous combustion of milk powder is the most widespread cause of fire and explosion in the spray dryer.

- It is a well-known fact that during the formation of oxidation products, excessive amount of heat is generated. The heat generation is greatly influenced by ambient temperature. Milk powder contains a number of oxidizable constituents such as fat, protein and lactose. These constituents undergo oxidation reaction when exposed to high temperature either during processing or storage.

- Any accumulation of material which is generating heat must lose that heat to the surroundings. If the deposit of powder is very small, all the heat generated can be lost and no hazard will take place. However, if the deposit is thick enough to prevent any significant heat dissipation from inside layers, a critical thickness may soon be reached, where the pited up heat can be a potential source of ignition. The critical temperature can vary with product layer thickness and can be lowered by the presence of oil. Ignition propagation time increases with decrease in critical temperature levels (Beever, 1985). Thick isolated layers of product can be created from non-operational conditions such as wet washing and incorrect start-up operation procedure.

- The composition and critical temperature of the product together with its location play a definite role in the mechanisms leading towards flame generation through spontaneous combustion (Beever, 1984).

- In the spray drying of milk products, it is found that the deposits of certain critical thickness can build up in the parts of the dryer, then, in principal, there is a risk of fire due to self-ignition.

- The time to ignition are likely to be of the order of an hour or less at 200°C and a few hours at 100°C (Beever, 1985).
- While dried milk is not highly explosive, the condition for dust explosion, apart from the ignition source, are almost always present in certain areas of a spray drying system during its normal operation (Beever & Growhurst, 1989).
- Self-ignition of deposits is known to be possible at the high operation temperature. At best, it can cause product degradation and discolouration and, at worst, it can provide the source of ignition for an explosion (Beever & Crowhurst, 1989).

2. Hot incandescent solid particles entering the dryer with the drying gas : Proper filtration of drying air is a must to avoid the red hot particles entering the drying chamber. Red hot particles can also be formed from the return of powder fines to the drying chamber for agglomeration, if there is a likelihood of powder entering or remaining in the air disperser.

3. Spark generation through friction : A mechanical spark so generated might be sufficient to ignite the dust cloud.

4. Electrical failure : An indirect effect of an electrical failure can be fire hazardous as well, causing undesirable friction of powder accumulation.

5. Static electricity discharge : When spraying the product in nozzles, the electrostatic charge of the droplet cloud can reach upto 20,000 V. This electrostatic discharge must inevitably occur in the spray dryer. However, the spark energy is low. Dried powder containing fat is easily inflammable by electrostatic discharge. Spark discharge through build-up of static electricity could, under certain circumstances, create explosions. Spray drier is normally a completely metallic structure and when earthed static electricity is nearly a problem.

6. Dust explosion : Milk powder in the spray dried plant is present in a physical state of finely distributed powder in the air in the form of a dust cloud (Skov, 1994). Certain conditions may cause the dust cloud to explode, such as :

- Ignition source of a certain strength. Most common ignition sources are open fire, hot surfaces and hot gases and burning materials.
- Oxygen exceeding a certain concentration.
- Burnable dust at a concentration within the explosion limit.

Safety Aspects

- The operating instructions for correct dryer operation should be closely followed.
- Inlet and Outlet temperature control during the operation and at the starting and shutting down stages should be rigorously practised.
- Air supply filters must be clean.
- All dryer parts and components should be cleaned regularly.

- Atomizer loading need to be controlled.
- Electrical parts, mechanical equipments, seals, bearings, packings etc., should be regularly inspected.
- Product discharge rate to establish any accumulation of product should be monitored.
- Atomizer oil leakage into the drying chamber must be rectified immediately.
- The temperature of the body of the exhaust air fan should be continuously measured and any difference in electric motor load should be signalised.
- Installation of vibration monitors on the atomizer shaft.

Preventive Measures

1. Air Supply

- Air must be drawn in from a dry and clean area and dust must be prevented from entering the air heater.
- Suction line must be fitted with a filter.
- Care must be taken that in no part of the plant a vacuum can be created, caused by the fans or by the rapid cooling of air when extinguishing.

2. Air Disperser

- Power must be prevented from depositing on air disperser as it is the hottest part of the drying chamber.
- A section of the roof of the drying chamber and the air disperser can be cooled by cooling air.
- Air carrying powder must be prevented from following back from the drying chamber into the air disperser, or to the heater, after plant shut down and cooling off period.

3. Atomization and Feeding System

- Atomization should be fully correct in order to avoid product deposits inside the drying chamber.
- The liquid distributor must be prevented from partial clogging.
- The correct consumption of the high pressure pump of the atomizer motor must be checked.
- Nozzle exteriors must be kept clean and cool at all times which can be achieved by shielding the nozzle and a minor low of cool and clean air around them.
- The spray pattern and any possible fouling of the nozzles should be visually detectable. T.V. monitors may be a simple solution, if the sensors are fitted in the upper part of the drying chamber.

- Concentration must not leak on the roof of the drying chamber and over the air disperser, else it may cause fire.

4. Drying Chamber

- Build-up of powder deposits is a problem in spray dryers. The problem is aggravated by fat filled powders. Installation of knocking hammers, vibrators, scrapers or air brooms are possible solutions to this problem.
- Cleaning must be carried out regularly for sticky powders.
- Product should not be fed unless the dryer is fully heated up.
- Temperature sensors should be installed atleast at the inlet and the exhaust of the dryer.
- The dryer can be equipped with fire detectors which are activated by any burning or glowing material. The purpose of the detectors is to activate the fire extinguishing system and to prevent fire or glowing material from spreading into the fluid bed and into cyclones and textile filters. The fire detectors are placed in the duct between the chamber and the fluid bed and in the duct between the chamber and the cylones.
- A fire extinguishing system should be provided consisting of sprinklers in the chamber ceiling (Fig. 5.25). The spray balls in the fluid-bed used for CIP should also be utilised in the fire extinguishing system.

- While using fluidized instantizers, care should be taken to minimize the appearance of large number of overrsized spheres as these help in spreading the fire rapidly.

5. Return of Fine Powder to the Drying Chamber

- Powder should be returned over the atomizer so that obstacles in the drying chamber are avoided.
- If the atomization is interrupted, powder return must be stopped immediately and automatically.

6. Final Drying and Cooling Beds

- Proper inspection and cleaning facilities should be provided for final drying and cooling bed.
- Air carrying powder must be prevented from falling through the air distributor or even from reaching the air heater.

7. Powder Separation and Transport

- Undesirable accumulation of powder due to blocking of cyclone or transport lines should be prevented.
- A cyclone that has been blocked for some time must not be emptied unless the plant has been stopped.

- Blocking of the pneumatic powder transport can be detected on the basis of the powder consumption or suction pressure of the fan.
- No powder should be allowed to accumulate in the bag filters.

8. High Pressure Relief Apertures

- Drying chambers with volume of 500-1000 m^3 should be equipped with relief areas of at least 2-4 m^3 (I.D.F. 1987).
- To avoid fire or secondary explosions within the building, the relief hatches should be ducted to the exterior of the building.

Action to be Taken in Case of Explosion or Fire

In any occurrence of the fire, the following steps must be undertaken as far as possible (Masters, 1985).

- Shut down fans and heater.
- Switch off the feed and pass off water to the atomizer.
- Stop powder valves.
- Do not enter the dryer.
- After the plant has cooled down, continue feeding water for a period of time.

DISCUSSION QUESTIONS

1. Define Roller dryer. Explain its important features.
2. Classify and explain drum dryers.
3. Define Spray dryer. Explain its merits and demerits.
4. Explain different components of a spray dryer.
5. Briefly describe powder conveying and recovering system.
6. What are the types of spray drying used in the dairy industry ? Explain.
7. Write brief notes on Tallform dryer and filtermat dryers.
8. How will you operate a powder plant ? Discuss.
9. Explain the major problems and remedy in a powder plant.
10. Discuss cleaning system of a spray dryer.
11. What are the possibilities of ignition and explosion in a powder plant ?
12. Explain the different safety measures in a powder plant.

6

Advances in Concentration and Drying Process

Dairy industry has gone through a mammoth change in processing technology, especially in terms of concentration and drying process. Membrane processing technology opened up a new dimension regarding separation process, in which it acts as a selective barrier to permit certain components and to retain certain other components. The catch words for today's dairy processing are Reverse Osmosis and Ultrafiltration. Both of these technologies brought a wave in every food and dairy processing. Reverse Osmosis (RO) is essentially a dewatering technique. In this process, hydraulic pressure is required to drive liquid through the membrane. RO came as a boon for concentration of whey, skim milk and milk. It has enormous potential for recovering lactose from whey, carbohydrate and protein from other sources.

Reverse Osmosis (RO)

RO process is also known as hyper filtration (H.F.). R.O. is essentially a dewatering technique. The commercial RO system normally uses the cellular acetate, polyamide and thin film composite (TFC) membranes, the latter having better properties for good applications. RO membranes have the smallest pore size in range of 1 to 10Å, needs the highest operating pressures compared to the other membrane separation processes. Normally, RO process operates at about 30 bar pressure for milk and milk by-products. RO membranes separate solutes with a molecular weight of approximately 150 Daltons and above. Hence, fat, proteins, lactose and all undissociated minerals are retained and concentrated by the membrane and only water and some ionized minerals are allowed to pass through (Cheryan, 1998). Energy requirements of membrane processes are very low compared with other removal process such as evaporation, freeze concentration, freeze drying etc. Table 6.1 and Table 6.2 compares the energy requirements of some common dewatering technique. There are about 150 reverse osmosis commercial plants worldwide in the food and dairy

Table 6.1

Energy Requirement of Some Dewatering Process

Process	Energy requirements KWh/1000 kg. water.
1. Evaporation (open pan)	626
2. Freezing	92
3. Vacuum Evaporation :	
3 or 4 effect	126-180
5 or 7 effect with MVR	37-53
4. RO (desalination)	4
5. RO milk	9-19

*Mechanical vapour recompression.

Source : Marshall, 1985.

Table 6.2

Comparison of the Energy Requirements for Evaporation and Reverse Osmosis for Water Removal

Concentration (Volume reduction)	Reverse Osmosis[a] Tubular Single Pass 30°C Electricity (KWh)	Reverse Osmosis[b] Flate Plate, Batch + internal Recirculation 10°C Electricity (KWh)	Evaporation[c]	
			Electricity	Gas d(m^3)
0-45% (TS[e] – 11%)	5	11	4.6	14
0-55% (TS – 13%)	—	13	4.6	14
0-66% (TS – 18%)	—	17	4.6	14
Cleaning	1.3	1.9	0.9	2.8

a Feed capacity 18,000 litres/hour, water removal 8100 litres/hour.

b. Feed capacity 7500 litres/hour, water removal 5000 litres/hour.

c. Five-stage evaporators, feed capacity 18,000 litres/hour, water removal 12,000 litres/hour.

d lm^3 of gas costs twice as much as 1 KWh of electricity

TS[e] = Total solids.

Source : Glover, F.A. (1985)

industry. Of these, over 100 plants are used for concentration of whey, skim milk and milk for various purposes (Pal and Cheryan, 1987).

Concentration of Skim Milk

In recent study, using CA membrane in spiral wound configuration, it was possible to achieve 3 : 1 concentration of Skim milk (9% TS to 27% TS) at a transmembrane pressure 400 psig (Pal and Cheryan, 1985). Higher concentration would substantially reduce the flux, thereby, negative the economic benefit of the process.

Since the maximum limits for concentration of skim milk by RO is upto a maximum of 3 fold (25-27% total solids), combination of RO process with single or multi effect evaporation is considered economical for higher concentrations (Stabite, 1983). The capital and operating costs of concentrating Skim milk from 8.8 to 25% solids by RO and then upto 45% solids by multi-stage evaporation were equivalent to those of a multi effect evaporation from 8.8 to 45% solids. However, energy requirements for the combination of RO and evaporation are about one-third less than that for multistage evaporation from 8.8 to 45% solids.

Benefit of RO Concentration Skim Milk

(1) Transportation of RO pre-concentrated milk to a main processing dairy reduces the hauling costs in proportion to concentration ratio.

(2) The use of about 2 fold concentrated Skim milk doubles the capacity of evaporators. (Donnelly *et al.*, 1974).

(3) The RO concentrated milk having upto 25% total solids can be used for manufacture of condensed and evaporated milk because the heat stability is not affected by RO process (Abbot *et al.*, 1979).

(4) The problem of cooked flavour and browning of condensed and evaporated milk can also be eliminated.

(5) It improves the product quality (Bennett, 1988).

Application of RO Concentrated Skim Milk

- Increasing the efficiency of evaporators.
- Product manufacturing, (ice-cream, condensed and evaporated milk, powder, yoghurt, Ymer, etc.).

Skim Milk Powder by RO Process

Skim milk powder can be manufactured either (a) directly from 4-fold RO concentrated milk .or (b) 2-2.5 fold RO concentrated milk followed by evaporation to about 47% solids (Abbot *et al.*, 1979). During the manufacture of powder, no problems are encounted and, thus, the use of RO can be freely interchanged with conventional evaporation (Dixon, 1985). It has been noted that the evaporation required less stream to attain the required concentration for the drier. Preheating of raw milk to a suitable tune and temperature combination is recommended before concentrations to ensure high bacteriological quality of Skim milk.

Benefits of RO Skim Milk Powder

- With respect to solubility and extent of whey protein denaturation, the RO process powders were slightly superior and there was no loss of vitamins. (Table 6.3).

Table 6.3

Vitamin Content (mg/g powder) of Powders Prepared from RO Retentates and by Conventional Evaporation.

Vitamin	Conventional evaporation	RO concentrate (2x)
Folic acid	0.46	0.45
Protein	0.19	0.19
Pantothenic acid	22.20	21.50
Nictomic acid	7.60	5.70
Riboflavin	12.60	13.90
Vitamin B_{12}	0.033	0.033
Vitamin B_1	2.22	2.10

Source : Abbot *et al.*, 1979.

Concentration of Whole Milk

Initially, RO process was mainly used for concentrating whey and Skim milk. Use of RO for concentration of whole milk is increasing rapidly because the presence of fat does not have a significant effect on the permeation rate and the quality of milk itself.

Problem and Solution for RO Concentration of Whole Milk

RO of raw milk will usually give concentrates of poor bacteriological quality and increase in free fat content. (Table 6.4).

Table 6.4

Effect of Reverse Osmosis Process (at 30°C) Using CA membrane on Bacteriological Quality and Free Fatty Acids Level of Whole Milk Concentration

Concentration ratio	Bacterial Count (Cfu/ML)		Free Fatty Acids (milli equiv./100 g. fat)	
	Raw Milk	Heated Milk	Raw Milk	Heated Milk
1.0	4.1×10^3	1.4×10^2	1.7	1.7
1.5	1.8×10^4	2.1×10^2	9.0	2.3
2.0	2.6×10^4	4.3×10^2	10.2	1.7
2.5	2.8×10^4	5.1×10^2	10.9	1.8

Source : Abbot *et al.*, 1979.

Milk heated to 80°C/20 sec. then cooked 30°C for RO.

Poor Bacteriological Quality

This problem can be overcome in this way :

(1) **Thermization of raw milk** : Thermization is an in-flow heat treatment of raw milk at 62-65°C for 10-20s, immediately followed by cooking (Van den Berg, 1984). Before R.O., thermization could be overcome by the problem of poor bacteriological quality (Abbot *et al.*, 1979).

(2) **Continuing developments** in membranes and improvements in the design of RO plants have also minimized bacteriological problems to a considerable extent (Marshall, 1985).

(3) **With the use of non-CA** membranes, bacterial growth problem can be reduced since higher operating temperature (50-55°C) can be used. This is particularly effective against heat sensitive psychrotrophic gram negative bacteria (Drew and Manners, 1985).

(4) **Bacterias continue** to increase in number in the RO concentrates during storage, unless it is pasteurized.

Damage of Fat Globules and Release of Free Fat

Other problem related to the quality of milk is the damage to fat globules resulting in release of free fat. This is due to the shearing action in the high pressure pumps and high operating pressure on the fat globules during concentration by RO. The release of free fat by active lipase enzyme results in off-flavours, particularly rancid, in milk and other dairy products.

This problem can be overcome by this way :

(1) Inactivation of the lipase enzyme by heating to about 60°C before concentration (thermization) would eliminate rancidity problems.

(2) Use of a restriction pipe of small diameter instead of pressure control valves in the retentate line would significantly reduce the amount of free fatty acids liberated when sub-pasteurized or thermized whole milk is concentrated by RO (Barbano *et al.*, 1983).

(3) Incorporating arrangements for cooling of concentrates below 10°C before pressure release would also reduce the liberation of free fatty acids (de Boer and Noony, 1980; Ver Streeg, 1985).

Whole Milk Powder from RO Concentrates

The removal of water from milk during the production of milk powder accounts for significant proportion of the cost of the powder.

Technology : Whole milk powder can be prepared using reverse osmosis as a first concentration step followed by either evaporation and spray drying or spray drying alone. Since milk is normally concentrated to 40-45% total solids before spray drying. RO cannot be used as a complete replacement of conventional evaporation. The reason for such a limitation of the process is that, the permeation of water through the membrane practically ceases beyond 3-fold (36-38% total solids) concentration. Major savings in energy can be achieved by concentrating whole milk to about 2.5 fold (30% TS) followed by concentrating to about 48%. TS using multistage evaporator, and finally spray drying. Heat treatment of milk before concentration has also been recommended with a view to inactivate the native lipoprotein lipase and improve subsequently the bacteriological quality of the product.

Properties of Whole Milk Powder

- No significant difference in the properties, viz., solubility, whey protein denaturation, powder morphology, and bacterial population appear between the powders made from milk concentrated by RO and those prepared by conventional evaporation (Abbot *et al.*, 1979).
- The free fat content of powders prepared from milk concentrated by RO are quite higher, the average values being 57 mg/g powder (range 27-61 mg) in comparison with 15 mg/g of conventionally prepared whole milk powder. (Abbot *et al.*, 1979). Such increases in the free fat content are undesirable as they are likely to enhance the rate of oxidation during storage of powder.

Homogenization of milk prior to concentration by RO could reduce the amount of free fat, but may cause heat instability in the concentrate before spray drying.

Freeze Concentration

Freeze concentration is a relatively new method for removing water from aqueous solutions, e.g., coffee, tea, fruit juice and milk.

Principle : Basic principle is cooling the solution to temperature below its freezing point to produce ice crystals, which can then be removed. The freeze concentration studies used a unique system to remove the ice crystals, leaving a solution with an increased TS content.

Single-stage freeze concentration : Experiment performed by Mil and Bouman (1990a) in a single-stage freeze concentration pilot plant with Skim milk, whole milk and diacetyl containing liquid. The data showed that concentration of TS upto 36, 38 and 49%, respectively, was possible. The product quality was good and the aroma retention also. Product losses were relatively low (<0.6% based on milk TS). However, extreme care must be taken regarding the cleaning and disinfection of the equipment.

Multi-stage counter-current process : This process largely overcome certain drawbacks of the single-stage process when operating in high concentration level. (Mil and Bouman 1990b).

Advantages : Freeze concentration offers several advantages in product quality, aroma retention and product yield.

The concentration procedure has no wide industrial application.

Freeze-drying

Freeze-drying has been used to produce very high quality powder. In this process, the product is deep-frozen and the water frozen in the product is evaporated under vacuum. It is also called sublimation drying. This ensures high product quality as, among other things, no protein denaturing takes place. The most important advantages of freeze drying is the greatest possible degree of preservation of the structure, the colour, the vitamin and amino acid content and flavour and odour of the freeze dried product. Freeze-drying is not widely used for the production of milk powder because of high energy demand.

DISCUSSION QUESTIONS

1. Define Reverse Osmosis (RO). What are the major benefits of RO ?
2. How will you manufacture Skimmed milk powder by RO ?
3. Explain process of manufacturing whole milk powder by reverse osmosis process.
4. Distinguish between freeze concentration and freeze-drying.
5. Explain different types of freeze concentration.

7

Defects : Causes and Prevention

Quality control is the major area of concern for every food and dairy industry around the globe. It synchronizes with the term 'defect' in food and dairy products. Eventually, absence of defects symbolizes the very presence of quality in the product. Products like SCM or USCM must conform with the physical, chemical and microbiological quality by eliminating its major and minor defects to stand tall in the food and dairy market. Any technological error in processing may lead to an undesirable product and consequently may incur heavy loss in economic terms. Organoleptic quality should take prime concern along with shelf-life of the products. By removing every defect the product can be smoothly marketed without any hassle.

Major Defects of Sweetened Condensed Milk

The major defects of sweetened condensed milk are categorized into two groups, viz., physico-chemical origin and microbial origin.

Physico-chemical Origin

- Sandiness
- Sugar sediment
- Age thickening
- Fat separation
- Browning
- Off flavour—Rancid.
- Bloats

Microbial Origin

- Mould button
- Bloating
- Thickening
- Off flavour—Rancid, stale

From among the defects, age thickening, sandiness and browning are most common and have serious implications in marketing of the product. Defects in sweetened condensed milk (SCM) may arise due to poor quality milk and faulty methods of manufacture and storage.

Physico-chemical Origin

Sandiness

(a) Due to lactose (6.12) crystal size : The relative smoothness to texture of SCM is determined mainly by the size of the lactose crystals it contains. The desired smoothness and velvety texture of SCM due to lactose crystals average size 10 microns or less. Sandiness is due to the presence of relatively large, coarse crystals of lactose.

Cause : Incorrect cooling and crystallization; excessive low temperature of storage of SCM.

Prevention : The size of the lactose crystals is controlled by the method and management of cooling the condensed milk.

(b) Due to sucrose : When the sucrose-in-water ratio of the finished product exceeds 64.5%, it approaches the status of a saturated solution of sucrose. This type of condensed milk stored in cold storage, the large sized crystals of sucrose is termed. This relatively large size crystal of sucrose give the condensed milk of a coarse and sandy texture.

Prevention : The size of the sucrose crystals is controlled by the method and management of cooling the condensed milk.

(c) Due to Rewarming of Too Cool Condensed Milk: When sweetened condensed milk become too cold for packaging and rewarm in presence of agitation, the sandiness developed.

Prevention : Warm the SCM without agitation.

Sugar sediment/sugar down

Deposits of sugar sediment on the bottom of can or barrel of SCM are commonly occurred. Such sedimentation consists predominantly of lactose crystals.

Cause : The cause of sugar deposits is due to the difference in specific gravity between the crystallized lactose and the remainder (colloidal protein) of condensed milk.

Factors Effecting Sugar Sediment/Sugar Down

- **Crystal size of lactose** : Condensed milk of normal viscosity in which the lactose crystals size is greater than 10 microns, sugar sedimentation occurs.
- **Homogenizing of milk** : It was shown that homogenizing increases the viscosity of milk. It decreases the tendency for the finished product to drop its lactose crystals. This effect increases with increasing homogenizing pressure.

- **Turning the cases during storage** : Turning the cases at specific intervals during prolong storage tends to minimize the amount of sugar sediment. However, turning the cases is not a preventive of lactose crystal deposit. In the case of milk with excessive crystal size or abnormal viscosity a sugar deposit will occur inspite of the turning precaution.

Prevention : Adequate viscosity and short storage can prevent the problem of sugar sediment.

Age Thickening

The term "age thickening" is used to distinguish the gradual increase in viscosity, due to physico-chemical changes, which occur during storage of condensed milk, from the thickening, sometimes caused by micro-organisms producing acid or rennet like enzymes. Age thickening affects all condensed milk. Its rate may be almost negligible to very rapid, so that after only a few weeks the milk is too thick to pour and is quite unfit for use.

Mechanism of Age Thickening Process

Sweetened condensed milk is essentially a suspension of lactose crystals and colloidal proteins in a saturated solution of lactose and sucrose. It belongs to the class of non-Newtonian liquids/fluids with the structure intermediate between the crystallized and coagulated.

Age thickening is believed to be caused by changes in the dispersed particles (protein) including interaction between these particles. Such changes/interactions are promoted or hampered depending upon the nature of protein colloid and characteristics of the dispersing medium in which the protein colloid is dispersed.

The viscosity of a concentrate is a function of the volume fraction of the dispersed particles (i.e., proteins). The volume fraction depends on the hydration of the protein, the protein composition and the protein content.

The solubility of αS_1, αS_2, and β-casein depends strongly on calcium concentration, pH and ionic strength. An increase in ionic strength and a drop in pH, which always occur on concentration of milk, lead to reduction in calcium binding to the protein, consequently result in a rise in voluminosity leading to an increase in viscosity.

Freshly prepared condensed milk is a system in which interaction between dispersed particles is relatively slight. During age thickening, the primary process involves increase in mass due to formation of aggregates which not only increases the particle size, but also change the particle shape and symmetry continued interaction among the protein micelles tend to develop a weak network which accounts for an increase in viscosity which is commonly termed as 'Age thickening'.

Factors Affecting Age Thickening

1. ***Quality and composition of milk*** : The variation in milk composition particularly in quality and quantity of proteins and minerals content influence the stability of milk and along with it the tendency for age thickening.

2. ***Fat/SNF ratio of the product*** : Fat acts as a softener and dilutent of protein suspension. It is capable

of retaining, a more syrupy, more pliant and more velvety texture. That is why a product made from Skim milk, incidentally has more pronounced tendency to age thickening than sweetened condensed whole milk.

3. *SNF-in-water ratio of the product* : Increase in SNF, particularly ratio of SNF-in-water increases the tendency to age thickening.

4. *Fore-warming treatment* : It is the most important step in the manufacture of SCM, from the stand point of control of age thickening. It affects fluidity stability as shown in Table 7.1.

Table 7.1

Effect of Forewarming on Age Thickening Tendency of S.C.M.

Temperature Regime	Fluidity Stability
60-65.5°C	Age thinning
65.5-75°C	Diminishes viscosity and age thickening tendency
80°C	Slight increase in age thickening tendency
85°C	Very definite effect
95-100°C	Most intense effect
Above boiling point	Decreases age thickening tendency
110-120°C	Very effective in controlling age thickening when milk is the least stable
Above 120°C	Age thinning
Forewarming by use of live steam"	Age thickening

Source : Hunziker (1949)

* Not recommended for case goods due to shelf-life consideration.

** Used in production of super-heated condensed skim milk.

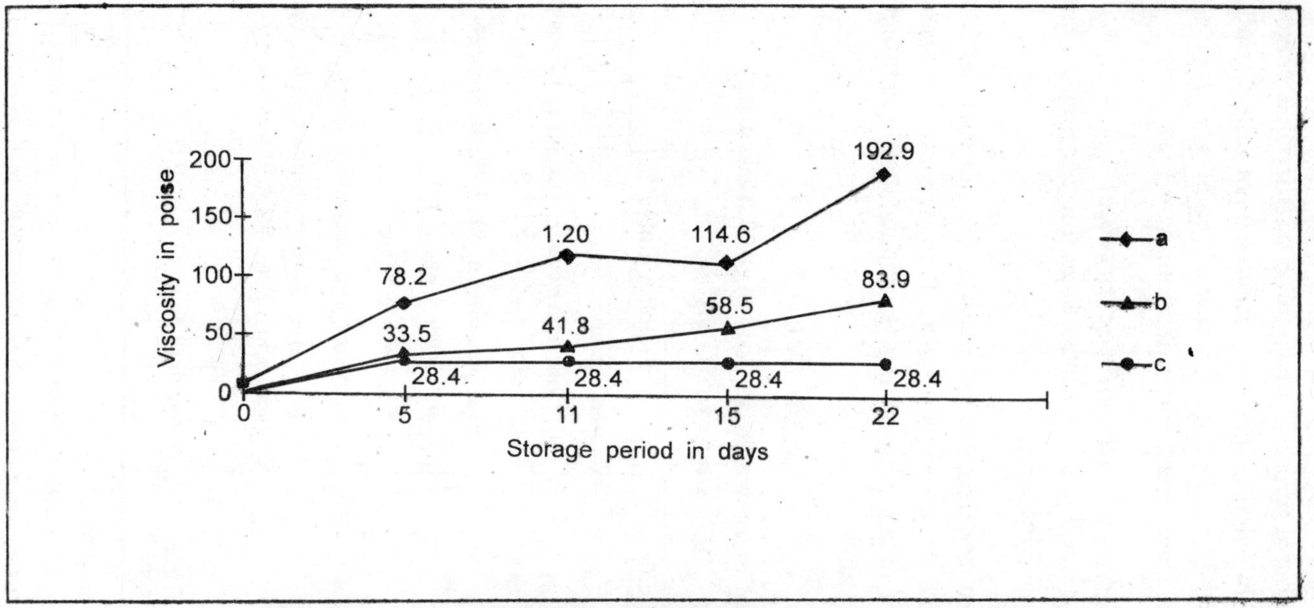

Fig. 7.1 Effect of Stage and Method of Adding Sugar on Age Thickening

(a) Sugar and milk forewarmed together, (b) sugar and milk forewarmed separately, cooled to 57°C, then mixed together before drawing into pan, (c) sugar added as syrup near the end of condensing period.

Source : Hunziker (1949)

5. *The amount of sugar added* : Sucrose-in-water ratio when increased results in reduced tendency for age thickening.

6. *Type of sugar added* : Impurities in sugar used and replacement of sucrose by dextrose in higher proportion (i.e., more than 25%) contribute to greater tendency for age thickening.

7. *Stage of sugar addition* : Addition of sugar to milk prior to forewarming is most conducive to age thickening. Adding sugar syrup at the end of the condensing period tends to cause age thinning, which may give rise to objectionable fat separation. Fig. 7.1 depicts influence of stage and method of adding sugar on age thickening.

8. *Temperature of condensing* : Condensing the milk at a higher pan temperature tends to unstablize the finished product and increases the tendency to age thickening. Fig. 7.2 shows influence of condensing temperature on rate of age thickening.

9. *Degree of concentration* : A product of high concentration has a high initial viscosity and tends to hasten age thickening. Too low a concentration, however, may cause fat separation and sugar deposit (Fig. 7.3).

10. *Temperature and storage time* : The storage temperature has a very marked effect on age thickening, thickening tendency increases with the storage temperature, particularly at temperature above 20°C.

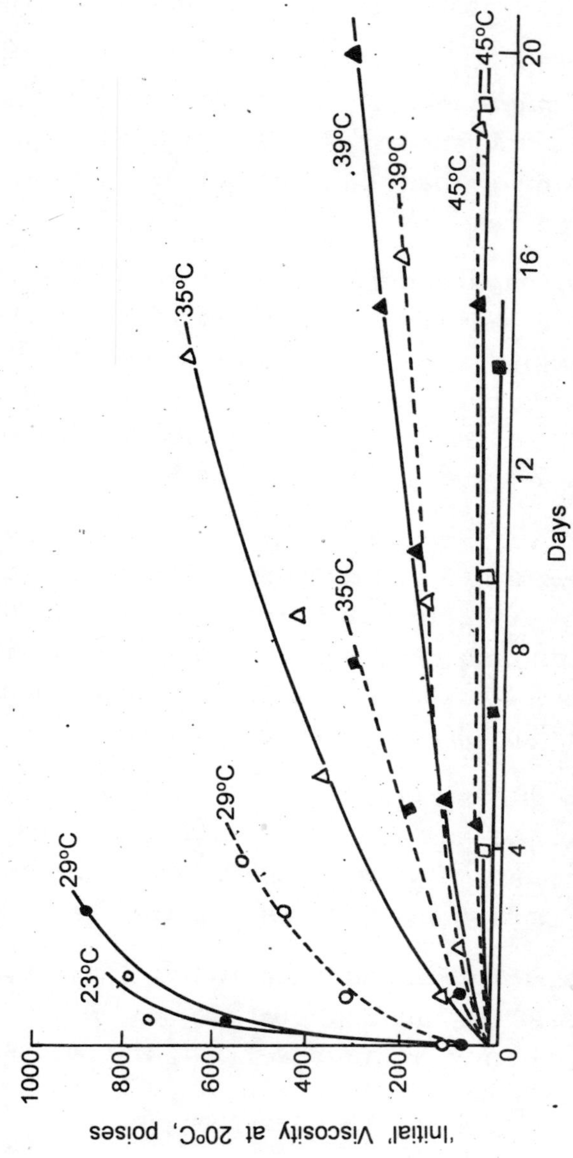

Full lines – milk not preheated; dotted lines – milk preheated (85°C, 10 min.)

Fig. 7.2 : Effect of Condensing Temperature on Rate of Age Thickening During Storage at 39°C

Source : Samel and Muers (1962 b)

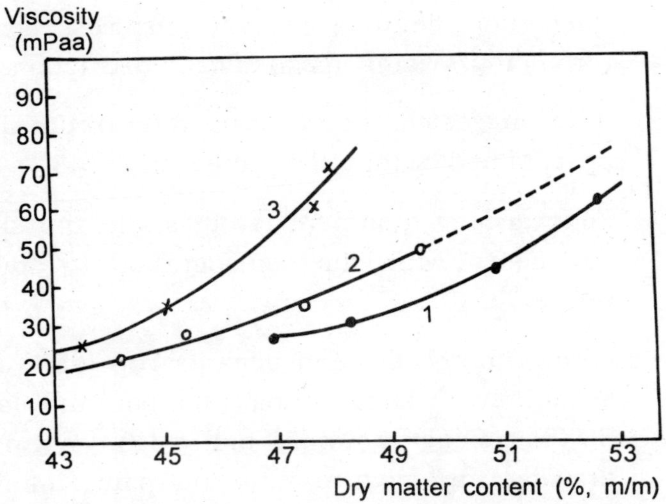

Preheat treatment of milk – 1 = 70°C, 10a; 2 = 85°C, 1 min. ; 3 = 95°C, 5 min. Shear rate = 392 S1

Fig. 7.3 : Viscosity of Skim Milk Concentrate as a Function of the Dry Matter Content

Source : Soneren, *et al.* (1982)

Prevention Measures of Age Thickening

1. The fresher and sweeter the fresh milk, the better chances for satisfactory stability and stability and keeping quality of the condensed milk.

2. pH adjustment of milk by 0.05-0.08 unit by addition of alkaline salt improves the storage stability of the finished product, particularly when unstability of the milk is related to acid reaction.

3. 1 SNF : 0.4090 fat has potential to lessen the interaction between protein particles, possibly through providing mechanical obstruction.

4. It is suggested that SNF-in-water ratio should be confined within the range of 44-48%.

5. Forewarming treatment is not static and should be chosen with due consideration to stabilize raw milk.

6. Forewarming milk and sugar syrup (65% conc.) separately, cooling to pan temperature (about 55°C), mixing the milk and sugar syrup and drawing the mixture into the pan diminishes the tendency to age thickening.

7. Age thickening can be prevented by removal of part of milk salts from milk by electrodialysis/ dialysis.

8. Addition of suitable stabilizing salts can prevent age thickening.

9. Stabilization through preliminary storage at below 20°C, and storage at low temperature are important measure of age thickening.

Fat Separation

Fat separation in sweetened condensed milk occurs infrequently. It is the result of abnormally low viscosity and thick body. In such milk, there is a definite tendency of some of the fat to rise to the surface forming a layer. The layer has a considerably deeper and clearer yellow colour than the body of the milk.

Cause : Adding sugar syrup at the end of the condensing period tends to cause age thinning, which may give rise to objectionable fat separation.

Prevention : Fat separation is readily prevented by method of manufacturers that are known to increase body viscosity.

Browning

The development of browning discolouration is usually accompanied by the development of a cooked flavour. But, discolouration of SCM is not as much of a problem as for evaporated condensed milk because lower total process and no sterilization takes place.

Cause : The immediate causes of the discolouration are the intense heat treatment of forewarming and high storage temperature.

Prevention : Optimum heat treatment during fore-warming and storage at low temperature below 10°C.

Rancid Flavour

Rancidity is not a frequent defect of SCM.

Cause : A rancid flavour is the consequence of insufficient heat treatment, where lipase deactivation was not complete or lipolytic microorganisms were not destroyed. The enzyme lipase and lipolytic microorganism hydrolyze the fat into glycerol and free fatty acids. This results objectionable bitter flavour and pungent, sharp unpleasant, rancid aroma.

Prevention

1. Optimum forewarming temperature.
2. Avoid leaks of raw milk into the batch during forewarming and between forewarming and vacuum pan.
3. Do not add sugar, nor remnant and rinsings of condensed milk to the fresh milk before or during the forewarming process.
4. Maintenance of sanitary condition.

Bloat (physico-chemical)

It is due to temperature shock to filled cans with non-hermetic seals.

Prevention : Bloat may be prevented by avoiding temperature shock to filled cans with non-hermetic seals.

Microbial Origin

Mold button

Mold buttons are lumps of variable size, cheese consistency and whitish yellow to reddish brown colour. They constitute firm, self-contained units that do not re-emulsify into the body of milk. The buttons themselves have a stale cheesy taste and they develop in the milk a stable odour. These buttons are found on the surface or in the subsurface layers of the product. They are entirely absent in freshly made SCM. Their appearance and size depends on the causative contamination, on the age of the milk and the temperature at which it is held. The mold grow until the availability of oxygen in head space is exhausted.

Causative organisms : *Aspergillus ripens, A. glaucus, Penicillium* spp., *Catenularia fulginea, Cladosporium* spp. certaina actinomyctes may also be involved in this defect.

Mechanism : Molds may develop and produce certain milk clotting enzymes which affect casein stability and form brown buttons. Buttons are composed of coagulated casein and mold mycelia usually coloured white to brown. Where the molds have lost their activity owing to lack of oxygen these 'buttons' still continue to grow owing to the activity of enzymes.

Factors Affecting Mold Button

- Contamination with molds after forewarming and their growth.
- Storage at high temperature.
- Presence of air in the head space cans.

Prevention

(i) Avoid contamination.

(ii) Store at low temperature (10°C or below)

(iii) Vacuum or gas packing and completely filling the containers.

(iv) Sanitation in all operations of manufacture and packaging that will prevent contamination.

(v) Inversion of stored cans at regular intervals help in reducing deterioration by molds.

Bacterial Thickening

Bacterial thickening is usually accompanied by a high bacterial count, a disagreeable cheesy odour and taste and an increase in acidity.

When the thickened product is diluted with water and heated, the curd separates. By this test, bacterial thickening can be distinguished from heat or age thickening.

Causative organism : *Micrococcus pyogenes, Staphylococcus aureus, Staph albus, Bacillus subtilis.* Certain yeasts may also be responsible for this defect.

Mechanism : The microorganism produce rennet—like enzyme which acts on milk protein and thickening of product takes place. When sugar contents are high, the growing condition of microorganism is rather unfavourable and thickening proceeds slowly.

Factors Affecting Bacterial Thickening

- Contamination and growth of microorganism which produce rennet-like enzyme.
- Sugar ratio is low.
- Storage at high temperature.

Prevention

- Avoid contamination.
- Store at low temperature (10°C/50°F).
- Use optimum sugar concentration.
- Improvement in plant sanitation.

Bloats/Gassy Fermentation

The gaseous fermenta-tion of sweetened condensed milk, resulting in the appearance of bloats, is one of the most disastrous defects. Gas may develop suddenly in the product during storage after 10 days to a few weeks. This spoiled product is commercially worthless and can be used only as animal feed.

The gas produced causes bulging of cans or blowing of cans or leakage and, in extreme cases, may break the cans or burst the barrels. The other signs are high acidity (unfit for consumption), lumpiness and darkening of interval surface of containers as well as the product in compact.

Causative organism : Bloat is due to the lactose and sucrose fermenting organisms (Fig. 7.4).

Yeast—*Torulopsis lactis condensi, Torulopsis globosa*

Bacteria—*Coliforms, Bacillus cereus, B. tutyricus, B. coagulans, Clostridium butyricum.*

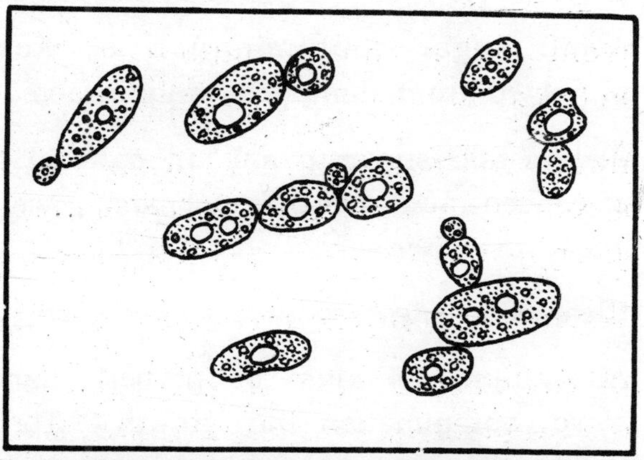

Fig. 7.4 Yeast cells causing gassy fermentation

Mechanisms : Sucrose is inverted by enzyme invertase generally produced by yeast cells. In this process, molecule of sucrose is converted to glucose and fructose.

$$\text{Sucrose} \xrightarrow{\text{Invertase}} \text{Glucose} + \text{Fructose}$$

The glucose and fructose molecules are fermented by the causative organisms with the subsequent production of gas like carbon dioxide or mixture of carbon dioxide and hydrogen and this causes the bulging and blowing of cans.

Factors Affecting Bloat

- The potential sources are the contaminated sugar or equipments.
- Causative organisms may draw oxygen from the head space of cans.

Prevention

- Avoid sugar contamination by keeping it protected from dampness and insects.
- Always add sugar to milk in pans in the form of concentrated syrup in hot boiling water (65%) sugar in water.
- Cleanse all sugar conveyers after each day's use.
- All equipments must be properly sanitized to avoid contamination and growth of Torula lactis condensi.

- Do not expose the product to air, fill the product as soon as possible after manufacture.
- Fill the cans fully and check for any leakage.

MAJOR DEFECTS IN EVAPORATED MILK

Microbial origin

Evaporated milk is generally considered to be sterile. Only in rare cases, bacteria are found in a can. Due to inadequate heat treatment, uneven distribution of heat in the sterilizer, thermophilic and heat resistant spore forming bacteria may survive and serve as spoilage producers.

Bloats/Gas Fermentation

Bloats caused by gaseous fermentation are commercially the most serious bacterial defect of evaporated milk. The gas causes bulging of cans.

Causative Organisms

The responsible organism is usually, though not always, a spore forming anaerobe like *Clostridium foetidum* or *Pectridium foetidum.* Aerobic species of B. coli group and Streptococcus also reported from the bloated evaporated milk.

Prevention

- Clean milk production on farm.
- Proper time–temperature combination for sterilization.

BACTERIAL COAGULATION

Bacterial coagulation is probably the most common defect in evaporated milk. It may be of a sweet or sour curd.

Sweet Coagulation

This type of defect is caused by sweet curding type spore forming bacteria. Besides coagulation, the other associated changes take place in the product. The coagulum formed can be digested, in which case a brownish liquid is formed. A disagreeable bitter taste is developed. It is due to production of proteolytic enzymes.

Sour Coagulation

It is due to acid forming bacteria milk coagulates owing to microbial acidity. Taste is somewhat sour and a little cheesy.

Causative organisms : *Bacillus coagulant* (sour coagulation), *Bacillus subtilis* (sweet curding), *B. cereus, B. calidolactis, B. megaterium, B. simplex, B. stearothemophilus.*

Factors Affecting

- High temperature of storage
- Inadequate heat treatment, lack of uniformity of heat distribution
- Contamination with the heat resistant organisms.
- Improper cooling.

Prevention

- Storage of the product should be done at cold temperature.
- Proper sterilization should be ensured.
- Adequate cooling should be observed.

Bitterness

Many instances of bacterial coagulum showed intensive bitterness. It is due to the breakdown and formation of peptones and other protein decomposition products and is, sometimes, accompanied by wheying off.

Causative organism : *B. amarus, B. panis.*

Prevention

Proper sterilization temperature and time combination should be used.

- Heat distribution should be uniform.

PHYSICO-CHEMICAL ORIGIN

Curdliness due to Heat Coagulation

The causes and prevention of curdliness are briefly summarized.

1. Curdliness of evaporated milk is due to low heat stability, in presence of the heat treatment used for sterilization.

2. High homogenizing pressure; inferior quality of raw milk also increases the tendency to curdliness.

Prevention : Curdliness prevented by the use of higher forewarming temperature or by the proper use of casein stabilizer.

Fat Separation

The milk in the top of the tin becomes increasingly richer in milk fat with progressive storage.

Cause : Inadequate homogenization pressure.

Prevention

- Effective homogenization (such as reducing the fat globules uniformly to a diameter of 2 microns).
- Moderate heat stability to give the evaporated milk full body and viscosity, which reduce fat separation in storage.

Mineral Deposition

Under certain conditions there is a tendency for whitish gritty deposit to appear in the bottom of the container cases. Citric acid is present in milk combined with calcium and magnesium. The higher concentration of the evaporated milk, the greater the tendency for appearance of mineral deposit during storage. The higher temperature storage, the larger the amount of deposit.

Prevention : Storage at low temperature and short storage can avoid this defects.

Browning

Colour change of evaporated milk is a consequence of lactose caramelization and Maillard's Maillard browning occurs when proteins interact with reducing sugars. This results in off-flavours and discolouration.

The colour of evaporated milk darkens with age, the rate of darkening increases with the storage temperature.

Prevention

1. The browning effect is much less intense in the case of HTST treatment during sterilization.
2. Storage at low temperature below 7°C and short storage help in preventing browning.

Cooked Flavour

It is the consequence of SH group activation during heat treatment. When milk is not heat treated, the SH groups of cystine in serum proteins and in particular, in β-dactoglobalin are masked and inactive. Thermal treatment produces denaturation and unwrapping of protein macromolecules. Simultaneous activation of the SH groups, and release of hydrogen sulfide.

Prevention : This flavour variation in evaporated milk is successfully minimized by HTST sterilization.

Age Thickening

Thickening, followed by gelation of evaporated milk during storage is a well known defects of this product. The main causes are excessive fore-warming temperature or excessive homogenization pressure and storage at high temperature.

Prevention

- Optimum fore-warming temperature and/or optimum homogenization pressure.
- Storage at low temperature below 7°C.

MAJOR DEFECTS IN WMP AND SMP

Stale Flavour

The development of stale flavour is with the age of milk powder due to physico-chemical changes of the protein constituents.

Prevention

- Low moisture content of milk powder, preferably below 3%.
- High quality, fresh sweet milk
- Airtight moisture proof package
- Low storage temperature 24°C.

Rancid Flavour

Rancidity in milk powder resembles the sour, irritating flavour and pungent rasping odour of butyric acid. It is the result of fat hydrolysis causing liberation of volatile fatty acids of which butyric acid is the most prominent. Hydrolysis producing rancidity is due to the presence of the fat splitting enzyme lipase. Low preheating temperature, does not inactivate the lipase.

Prevention : Rancidity in milk powder is readily and dependently prevented by heat treatment in the process of manufacture that will destroy or permanently inactivated the enzyme lipase.

Fishy Flavour

Fishy flavour is relatively uncommon defect of dried milk. When it does occur, it is usually associated with high moisture. Lecithin hydrolysis appears to be the most likely cause. Oxidation of the unsaturated fats in the presence of moisture and protein substances has also been suggested as a probable cause.

Prevention : Reduction of the moisture content to below 3.5% prevents the development of fishy flavour in dried milk.

Tallowy/Oxidized Flavour

Tallowy flavour is most common and serious flavour defects of whole milk powder. Whole milk powder containing milk fat and anti-oxidation of fat causes the tallowiness of product.

Prevention

- Use of fresh, sweet milk only.
- Avoid contact of milk with copper and iron.
- Keep the equipment in sanitary condition.
- Remove the powder from drying chamber promptly.
- Low moisture content is essential.
- Gas packing helps to prevent tallowiness.

PROTEIN-LACTOSE DETERIORATION

In the presence of high moisture in powder during storage, a chain of chemical reaction involving lactose and protein takes place which results in the deterioration in powder quality. Initially, lactose absorbs moisture and changes its form, from amorphous to crystalline. The powder may become sticky and lumpy and its apparent moisture content may decrease. Next, the casein progressively becomes denatured and insoluble. At the same time, protein combines with lactose to form a protein-lactose compound. Lastly, this compound decomposes with absorption of oxygen, to form brownish coloured products and substances of unpleasant stale, cardboard or gluey flavour. All these defects normally occur together and are commonest in skim milk powder, which is not hermetically packed and absorbs moisture during storage.

Prevention

- Gas packing
- Low storage temperature.
- Moisture content of the powder is controlled well within 5%.

DISCUSSION QUESTIONS

1. Name the common defects in SCM. Why is defect elimination must for dairy industry ?
2. Explain the physico-chemical defects in sweetened condensed milk.
3. What are the factors responsible for sugar related defect in SCM ?
4. Define age thickening. How will you prevent it ?
5. What is Mold Button ? Describe the cause and prevention of Mold Button in SCM.

8

Baby Foods, Malted Milk Powder and Dairy Whiteners

Throughout the ages, human milk is unquestionably considered as the best and ideal food for the infants. But, many a time, due to some physiological disorder, one has to get an appropriate alternative to cope up with this problem. A country like India, where 30-50% children are suffering from malnutrition, infant milk food or the baby food may be a suitable answer at present. Other variations may be malted milk powder to boost and energise the children with a readily digestible form of protein and sugar. These are having pleasant flavour and palatable in nature. Dairy whiteners are the new arrival for dairy and its allied industries, commonly used forms are tea or coffee whiteners which may be available as dry sodium caseinate and a supplement food for infants and diabetic patients.

INFANT MILK FORMULAS

Infant milk formulas are designed as a substitute for human milk when lactation failure, insufficient milk secretion and where mothers suffer from transmittable diseases. Usually, cow's milk is used with its composition modified to be as close to human milk as possible. The manufacture of infant milk formulas based on cow's milk can be traced to the beginning of twentieth century. Since then, their production has increased steadily. In India, investigations were jointly undertaken by Central Food Technological Research Institute, Mysore and AMUL Dairy, Anand in 1957 and a standardized process was evolved for the commercial production of infant food from buffalo milk in 1960 by the Roller drying process, followed by spray drying process in 1967 and became a marketable commodity. The dried milk industry manufactures nearly 1,85,000 tonnes of infant milk formula in India, representing approximately 3.8% of total milk production (Thompkinson *et al.,* 2005).

Definition

Infant milk food (baby food) means the product obtained by drying cow or buffalo milk or a combination thereof or by drying standardized milk with the specific addition of carbohydrates (cane sugar, dextrose and dextrines, maltose or lactose); iron, salts and vitamins. It shall be free from starch and antioxidants.

Legal Standards

Composition standards for Infant food are given in Tables 8.1, 8.2, 8.3 and 8.4

Table 8.1

Compositional Standards for Infant Foods#

Attributes	IS : 1547		
	1960	1968	1985
Moisture % (max)	2.8	4.5	4.5
Total milk protein % (min.)	20.0	20.0	12.0
Milk fat %	18.0-28.0	18.0-28.0	18.0*
Carbohydrates % (min.)	35.0	35.0	—
Total ash % (max.)	8.5	8.5	8.5
Vit. A IU/100 g (min.)	1500	1500	1050
Vit. D IU/100 g	400-800	400-800	180
Iron mg/100 g (min.)	4.0	4.0	5.0
Bacterial Count Perg (max.)	50,000	50,000	40,000
Coliform Count Perg (max)	10	10	Nil

* Milk fat content should not be less than 12%

Indian Standards.

Table 8.2

Revised Compositional Standard for Infant food*

Attributes	IS : 11156 1985	PFA 1989	PFA 1991
Moisture % (max)	4.5	4.5	4.5
Milk protein % (max)	12.0	12.0	10.0-16.0++
Total fat % (min.)	18.0	18.0*	18.0**
Total ash % (max.)	8.5	8.5	8.5
Linoleates g/100g	1.398	1.398	1.398
Vit. A IU/100 g (min.)	1050	1050	1050
Vit. D IU/100 g	350	350	350
Vit. E IU/g linoleate	0.70	0.70	0.70
Iron mg/100 g (min.)	5.0	5.0	5.0
Bacterial Count/g (max.)	40,000	40,000	40,000
Coliform count/g (max.)	Nil	Nil	Nil

Indian Standard, PFA Standards.

* Milk fat content should not be less than 12%

** Restriction from min. 12% milk fat deleted.

++ 10-14% for starter formula & 10-16% for follow-up formula.

Table 8.3

Minor Constituents for Infant Food*

Constituents	(1985)[1] min/100g	(1992)[2] min/100g
1	2	3
Thiamine (mg)	185.0	185.0
Nicotinamide (mg)	1160.0	1160.0
Riboflavin (mg)	275.0	275.0
Vit. B_6 (mg)	160.0	160.0
Vit. B_{12} (mg)	0.7	0.7
Folic Acid (mg)	20.0	20.0
Biotin (mg)	7.0	7.0
Vit. K (mg)	18.0	18.0
Copper (mg)	280.0	280.0
Iodine (mg)	20.0	20.0
Manganese (mg)	20.0	20.0
Pantothenic acid (mg)	1.4	1.4
Vit. C (mg)	35.0	35.0
Sodium (mg)	—	90.0
Potassium (mg)	—	370.0
Chloride (mg)	—	250.0
Magnesium (mg)	—	22.0
Calcium (mg)	—	230.0
Choline (mg)	—	32.0
Phosphorus (mg.)	—	115.0
Zinc (mg)	2.5	2.5

Indian Standard

1. Optimal addition of listed micro nutrients
2. Compulsory addition of micronutrients.

Table 8.4

Minor Constituents of Infant formulas (per 100 Kcal)

Nutrient (Mins. std.)	Codex 1981	PFA 1998
Vitamins		
A (iv)	250.0	239.7
D (iv)	80.0	79.9
K (mg)	4.0	4.1
E (IU)	0.7	0.7
C (ascorbic acid) (mg)	8.0	7.99
B_1 (Thiamine) (mg)	40.0	42.2
B_2 (riboflavin) (mg)	62.0	62.8
B_6 (pyridoxine) (mg)	36.0	36.5
B_{12} (mg)	0.15	0.16
Nicotinamide (mg)	260.0	264.0
Folic acid (mg)	5.0	4.6
Pantothenic acid (mg)	300.0	320.0
Biotin (mg)	1.5	1.6
Minerals :		
Calcium (mg)	50.0	52.5
Phosphorous (mg)	25.0	26.2
Magnesium (mg)	6.0	5.0
Iron (mg.)	1.0	1.1
Iodine (mg)	5.0	4.5
Zinc (mg)	0.55	0.57
Copper (mg)	62.0	63.9
Manganese (mg)	5.0	4.5
Sodium (mg)	20.0	20.5
Potassium (mg)	80.0	84.4
Chloride (mg)	55.0	57.1
Choline (mg)	7.0	7.3

Source : Codex (1981); PFA (1999)

Composition

Gross composition of selected brands of Infant formula marketed in India is given in Table 8.5. Commercially available infant formulae generally lack in bioprotective factors, firstly, because bovine milk is inherently deficient in bio-immune factors and, secondly, due to thermal processing treatments that tend to further reduce their level.

Table 8.5

Gross Composition of Selected Brands of Infant Formula Marketed in India

Brand	Constituents g/100g powder					
	Protein	Fat	Carbohy-drates	Minerals	Moisture	Energy (K Cal)
Amul	22.0	18.0	50.0	6.8	3.5	450
Lactogen Follow-up Formula	21.6	19.0	51.6	4.8	3.0	464
Lactogen Starter Formula	16.1	24.2	53.1	3.6	3.0	495
Glaxo Sunshine	22.0	18.0	50.0	6.5	3.5	450
Lactodex	14.5	6.5	72.9	3.5	3.0	408
Farex	15.5	7.5	70.5	3.5	3.0	408
Sapan	21.0	18.0	52.4	5.8	3.5	458
Angel	21.0	18.0	52.4	5.6	3.0	456
Madhu	22.0	18.0	50.0	6.5	3.5	450
Raptakos	12.0	20.0	69.0	2.8	3.0	484
Milk Care	12.0	18.0	51.4	5.6	3.0	458
Lever Spray	22.0	18.0	51.0	5.8	3.2	454

Manufacture Technology of Infant Milk Formula/Food

There are two procedures prevailing in the manufacture formula. One is 'dry procedure' in which all ingredients are blended in dry form; the other is the 'wet procedure' in which wet or liquid ingredients are mixed, prior to drying. The 'wet procedure' consists of the following operations : reception and selection of raw material (Skim milk) clarification, deaeration, separation, pasteurization, blending with oil and other components (emulsifiers, stabilizers, fat-soluble vitamins), mixing homogenization, adding of water soluble vitamins and minerals and drying (Fig. 8.1).

There is a combination of these two processes in which water-soluble components are added to the milk before drying, while less soluble components are added in a dry form of the blend after drying. The 'wet procedure' provide best mixing resulting in a sterile product, while the 'dry procedure' is less expensive for the investment and operation. The combined method has the advantages of both and is most commonly applied.

The Manufacturing Steps

Milk reception : Only high grade raw milk should be accepted. Raw fresh milk must be of low acidity and have a low total microorganism count. If skim milk powder is used as raw material, it is reconstituted before being added to the blend during mixing operation.

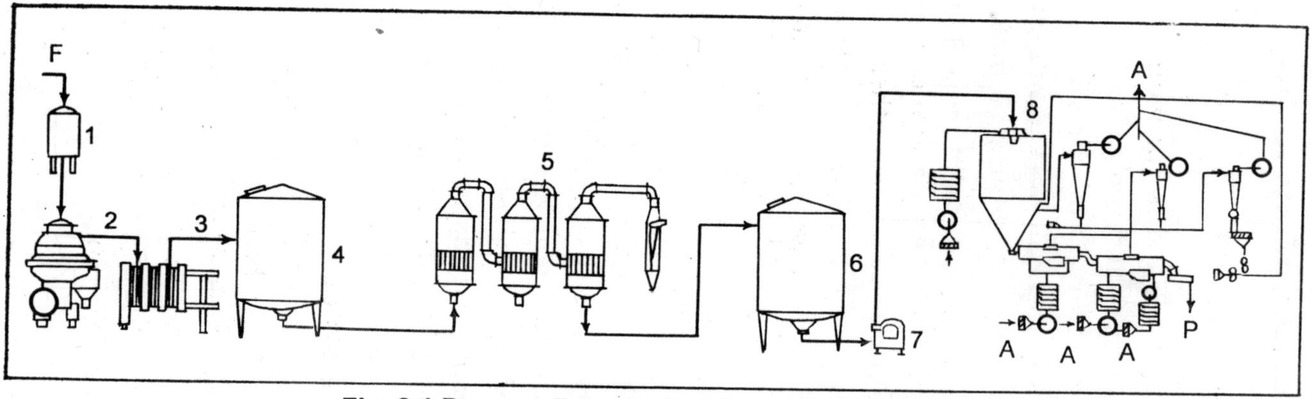

Fig. 8.1 Process line of infant formula production

1. Receitpion tank, 2. Separator, 3. Pasteurizer, 4. tank, 5. Evaporator, 6. Tank, 7. Homogenizer, 8. Spray drier with fluid beds and cyclones. F = Feed, A = Air, R = Product.

Clarification : Clarification is formed to remove impurities from the milk.

Standardization : It is done to conform to legal standard. The standardized milk may contain 2.9-3.05% fat and 9.6% SNF. Phosphate buffer solution is added to react with Ca and reduce the curd tension of milk. Ferrous sulphate is added to provide Fe and sugar to increase the palatability.

Forewarming : The milk is preheated to 80°C/no hold for destruction of pathogens, other harmful microorganisms and to inactivate enzyme. The choice of temperature employed for forewarming tends to vary from manufacturer to manufacturer depending upon the microbial quality of milk as well as reconstitution and sensory characteristics desired in the final product.

Evaporation : It is necessary if the formulas are composed in such a way as to contain less than 40-80% dry matter after blending. Evaporation is performed in multistage vacuum evaporators. In this phase of processing, most of the necessary components are added to Skim milk, vegetable oils (a corresponding mixture of modified fat), demineralized whey, fat-soluble vitamins, emulsifier and stabilizer.

Homogenization : To reduce size of fat globules.

Spray drying : It is usually performed in infant formula production. Infant formula contains high fat and lactose content. Lactose could cause the powder to stick in the processing machinery, pumps and pipelines. To avoid this, the following drying conditions must be met : relatively low temperature of inlet air (160-180°C), low dry matter of feeding concentrate (upto 45%) and special construction of the chamber with cooling during drying. The temperature of outlet air is 90°C or lower in multistage drying. Drying infant formula production is usually accomplished in two or three stages.

Comparison between Human and Bovine Milk

A comparison of the average composition in human and cow's milk (Table 8.6) shows basic differences in the quality of lactose, proteins, minerals and fat. Differences in the composition of proteins, fats and minerals are also apparent (Tables 8.7, 8.8, 8.9, 8.10). Cow's milk has 3.5 times more proteins than human milk. Moreover, cow's milk proteins have 80% casein and 20% whey protein while human milk proteins are composed of 20% casein and 80% whey proteins.

Table 8.6

Composition of Human and Cow's milk

Component (%)	Human	Cow's Milk
Water	87.43	86.61
Fat	3.75	4.14
Protein	1.63	3.58
Lactose	6.98	4.96
Ash	0.21	0.71
Non-fat solids	8.82	9.25
Total solids	12.57	13.39

Source : Gurr, 1981

Table 8.7

Protein Composition of Human and Cow's Milk

Protein	Human milk		Cow's Milk	
	g/100g	% of Total	g/100	% of total
Total	3.30	100	0.88	100
Casein	2.60	79	0.31	35
Serum protein	0.70	21	0.57	65
α-Lactabumin	0.12	3.5	0.15	17
β-Lactoglobulin	0.30	9.0	0.00	—
Lactoferrin	Traces	—	0.15	17
Serum albumin	0.03	1.0	0.05	6
Lysozym	Traces	1.0	0.05	6
Immunoglobulin	0.10	3.0	0.10	11
Others	0.15	4.5	0.07	8

Source : Packard (1982)

Table 8.8

Fatty Acid Composition of Human and Cow's Milk Fat

Fattyacids	Human milk (%)	Cow's milk (%)
1	2	3
Saturated		
Butyric (4 : 0)	—b	3.5
Caproic (6 : 0)	—	1.9
Caprylic (8 : 0)	—b	1.3
Capric (10 : 0)	1.4	2.5
Lauric (12 : 0)	6.2	2.8

contd.

1	2	3
Myristic (14 : 0)	7.8	10.7
Palmitic (16 : 0)	22.1	27.8
Stearic (18 : 0)	6.7	12.6
Total	48.2	65.6
Monounsaturated		
Palmitoleic (16 : 1)	3.1	2.5
Oleic (18 : 1)	35.5	26.5
Gadoleic (20 : 1)	0.96	Trace
Cetoleic (22 : 1)	Trace	Trace
Total	39.8	30.3
Polyunsaturated		
Linoleic (18 : 2)	8.9	2.5
Linolenic (18 : 3)	1.2	1.6
Parinaric (18 : 4)	—b	Trace
Arachidonic (20 : 4)	0.72	Trace
Eicosapentenoic (20 : 5)	Trace	Trace
Docosa pentenoic (22 : 5)	Trace	Trace
Docosahexenoic (22 : 6)	Trace	Trace
Total	12.0	4.1

Source : Packard (1982); US Department of Agriculture (1976).

a. Number in parentheses following the fatty acid name indicate the number of carton atoms and number of double bonds, in that order. Common or scientific names are given for case of discussion in the text. However, unsaturated fatty acids include both positional and geometric isomers.

b. Presence of this fatty acid is uncertain.

Table 8.9

Concentrations of Mineral Elements in Human and Cow's Milk

Mineral	Human Milk (mg/100g)	Cow's milk (mg/100 g)
Calcium	32	119
Phosphorus	14	93
Magnesium	3.0	13
Iron	0.03	0.05
Potassium	51	152
Sodium	17	49
Zinc	0.17	0.38

Source : Packard (1982).

Table 8.10

Vitamin Content of Human Milk and Cow's Milk (amount per 100 ml.)

Vitamin	Human Milk Colostrum		Cow's Milk Colostrum	
	day 1-5	day 6-10	Mature	Pure
1	2	3	4	5
Water soluble				
Ascorbic acid (mg)	4.4	5.4	4.3	1.6
Thiamine (mg)	15	6	16	42
Riboflavin (mg)	29.6	33.2	42.6	157
Niacin (mg)	75	175	172	85
Pyridoxine (mg)	—	—	11	48
Pantothenic acid (mg)	183	288	196	350
Vitamin B_{12} (mg)	0.045	0.036	Trace	0.56

(Contd.)

1	2	3	4	5
Biotin (mg)	0.1	0.4	0.4	3.5
Choline (mg)	—	—	9	13
Inositol (mg)	—	—	39	13
Fat-soluble				
Vitamin A (mg)	89	88	53	34
Total Carotenoids (mg)	112	38	27	38
Vitamin D (IU)	—	—	0.42	2.36
Vitamin E (mg)	1.28	1.32	0.56	0.06
Vitamin K (mg)	—	—	1.5	5.8

Source : Causeret (1977)

RECENT STUDIES ON DIFFERENT MODIFICATION OF BOVINE MILK FOR THE DEVELOPMENT OF INFANT FORMULA

Modification of Buffalo Milk Lipid

Infant milk food formulations that are presently marketed in India derive lipid exclusively from bovine milk. With a view to provide adequate linoleates and balance of saturated to unsaturated fatty acid (P/S), it is desirable to modify bovine milk lipid (Gurr, 1981). The Indian Standard (ISI, 1985) permits incorporation of upto 6% of vegetable oils to the infant formulation. Vegetable oils rich in PUFA have been considered most ideal for the modification of bovine milk lipids (A-bd-EL-Salam *et al.*, 1973, Ganguli and Kuchroo 1979, Mettler, 1980).

The study made by Thompkinson and Mathur (1987) to approach the level of human milk lipid. The requirement of fat in the formulation was adjusted to 20%. This 20% fat was comprising 63% milk fat and 37% of groundnut oil. This proportion of milk fat and vegetable oil works out to be 12.6% and 7.4% of total fat, respectively. On the basis standard for infant formula, the fat phase should contain not less than 12% milk fat. The fat blend developed by Thompkinson and Mathur (1987) contains 12.6% milk fat, which is in accordance with the prescribed compositional requirements. Table 8.8 shows the fatty acid % distributed of blend in comparison to human milk lipid.

Pre-term Infant Formula

Considering the special nutritional requirements of low birth weight and pre-term infant, a spray dried product was developed at NDRI laboratory for the dietary management for this group of infant (Rao and Mathur, 1987 a, b). Infant formula intended for the dietary management of pre-term subjects was so prepared as to contain protein 15%, milk fat 14%, groundnut oil 12%, lactose 19%, maltodextrin 28%, sucrose 5%, minerals 3.5%, moisture 3.5% (Rao and Mathur, 1987). The product containing 15% protein and 26 per cent fat with a protein fat : Carbohydrate ratio 1 : 1.75 : 3.5 was best suited for pre-term infants as per their nutritional requirements. Retention of vitamin activity was 72.3 per cent for vitamin A, 86.1% for vitamin D and 83.3 per cent for vitamin E.

Fortification with Bifidobacteria

The intestinal tract of foetus is generally free from microorganisms but at the time of birth, a massive invasion of microorganism takes place from birth canal and faeces resulting in a very quick establishment of microflora in the gastro intestinal tract of the infant. Human milk contains bifidus growth stimulating factors (BSF) which promote and ultimately lead to dominance of *Bifidobacterium bifidum* in the intestinal tract amongst breast fed infants. In absence of BSF, as in case of bottle fed infants, *Escherchia* sp. predominates (Gyorgy, 1953). Presence of *Bifidus* flora in concert with numerous host resistance factors present in breast milk provides protection against enteric infection (Goldman and Smith, 1973). Additionally bifidobacteria synthesis 'B complex' vitamin (Keele and Neil, 1966). Create acidic environment which promotes absorption of nutritionally superior minerals and exerts detoxification influence in chronic liver diseases (Meyer, 1969).

Infant formula prepared from buffalo Skim milk and rennet whey with added viable organisms of *Bifidobacterium bifidum* has been shown to process bifidus activity (BA) comparable to that of breast milk (Pawha, 1982). The preparation of bifidobacterial starters and their use in fermented infant formula has been described and patented. Pahwa (1982), at NDRI, Karnal, compared the effects of feeding a non-fermented infant formula fortified with a spray dried concentrate of bifidobacteria (final concentration of 5.73 log cfu/ml) with those of human milk and conventional infant formula. Fortification decreased the fecal pH and increased the bifidobacteria to level obtained in human milk. The fecal coliform counts decreased from 7.5 to 7.09 cfu/ml as compared to 6.6 for human milk. The retention of calcium, phosphorus and iron also significantly increased, most probably because of the low intestinal pH.

Modification of Buffalo Milk Casein

Kuchorro and Ganguli (1980) prepared a humanized infant formula from buffalo milk containing low levels of the antigenic as casein. Electrodialyzed Skim milk was subjected to protein hydrolysis with trypsin, pasteurized and fortified with lactose and vegetable oils. After homogenization, vitamin mix was added and spray dried. Commercial hypoallesgenic infant formula are Nentramagin and Podegestamil of Mead Johnson laboratories, contain casein hydrolyzates.

Low-lactose Infant Formula

A process was developed for the manufacture of low lactose infant formula for the dietary management of infants suffering from lactose mal-absorption (Paul and Mathur, 1993a, b). They recommended a 35 per cent level of lactose hydrolysis in a mixture of buffalo Skim milk and whey as base (Table 8.11). The physico-chemical and reconsti-tutional attributes remain satisfactory and the available lysine content decreased only marginally during storage. The spray dried product containing viable cells of *B. bifidum* could further augment the physico-logical and immunological systems in infants to combat lactose insufficiency and entire infections.

Table 8.11

Compositional Characteristics of Low-lactose and Pre-term Infant Formulae

Constituent	Low-lactose formulae	Pre-term formula
Protein (%)	13.2	15.0
Fat (%)	20.0	26.0
Carbohydrate (%)	61.0*	52.0
Ash (%)	2.6	3.5
Moisture (%)	3.5	3.5
Ca (mg/100 g)	995	976
P (mg/100 g)	496	488
K (mg/100 g)	880	920
Na (mg/100 g)	246	270
Fe (mg/100 g)	5	10

* Lactose content : 18%

Source : Paul (1984); Rao (1985)

Low lactose formula based on milk protein coprecipitate or sodium caseinate can be manufactured as shown in Fig. 8.2 (Rao, 1981). This product was found to have excellent PER (2.7 for the caseinate–formulation and upto 3.3 for the coprecipitate formulations).

Supplementation with Lysozyme

Breast fed babies fall ill less often than babies on diets based on cow milk, indicating to the presence of antimicrobial substances in mother's milk. The substances in human milk which have an antimicrobial effect are immunoglobulins, lactoferrin, lysozyme lactoperoxidase and vitamin B_{12} binding protein. Lysozyme is involved in the

Fig. 8.2 Flow diagram for the manufacture of a low lactose infant food.

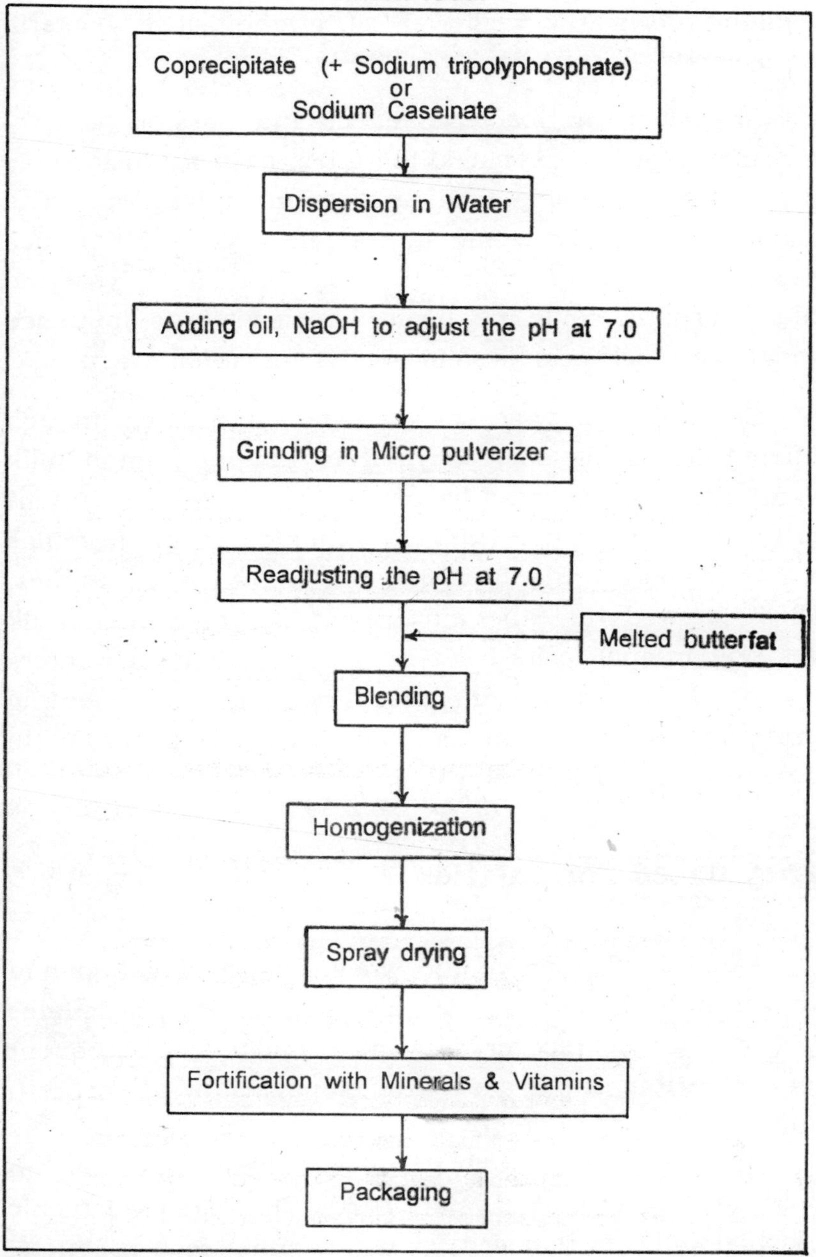

Source : Rao 1981.

milk immune system. Lysozyme has a direct effect by breaking down the cell wall of gram-positive bacteria (Kansal, 1992).

In view of antibacterial effect and possibly general immune system of neonates, there has been a considerable interest in developing food application of lysozyme. The concentration of lysozyme in human milk is 300 times greater than bovine milk. The incorporation of lysozyme infant formula could be useful in view of its influence under the ecological system of the intestinal tract.

Egg white lysozyme has a similar composition (though different antigenic specificity), than that of human milk lysozyme, and is readily available for use in infant formulas. A Japanese firm has patented infant formulas containing egg white lysozyme, on the basis of studies, which showed that the biological value of human milk was proportional to its lysozyme content. Russian workers developed and patented procedures for supplementing fermented infant formulas with lysozyme in order to increase their biological value.

Whey Based Formulation

After clarification of the whey by centrifugation and concentration by reverse osmosis (RO), lactose precipitates out and can be recovered by centrifugation. Subsequent gel filtration facilitates the separation of protein components and removal of ash β-lactoglobulin is antigenically most reactive and is often implicated in child's allergy to milk (Lyster, 1970). The highly allergenic β-lactoglobulin is removed for use in other food products, and the remaining whey proteins are blended to the ratio of human milk. The blend is then evaporated and spray dried. A β-lactoglo-bulin free whey protein concentrate could also be prepared by a method that involves selective precipitation of the proteins with ferric chloride and removal of the iron by ion exchange. Composition of whey protein concentrate base infant formula dry mix is given in Table 8.12.

Table 8.12

Composition of Whey Protein Concentrates (WPC) Based Infant Formula Dry Mix

Ingredients	Level (%)
Lactose	37.00
Fat blend	27.00*
WPC-34	18.5
SMP	16.00
Vitamins and minerals	1.00*
Lacithin	0.50
Water	To 100.00
Total	100.00

Source : www.usdec.org
WPC—Whey protein concentrate
SMP Skim milk powder
* vitamins, minerals and fat contents to be adjusted to meet the legal standards.

Addition of Lactobacilli

A number of other closely related lactobacilli, which inhabit the human intestinal tract, synthesize organic acids (acetic, benzoic, lactic) hydrogen peroxide, B-vitamins and enzymes, which may provide nutritional and/or therapeutic benefits. Langhendries *et al.* (1995) reported that the addition of lactobacilli to infant formula resulted in a gut ecology similar to those of breast-fed infants.

Buffalo Milk Protein

Joglekar (1984) prepared an infant formula wherein the protein was exclusively derived from buffalo milk to the extent that an adequate amount of limiting essential amino acid could be supplied at comparable levels of human milk (Table 8.13).

Table 8.13

Amino Acid Profiles of Buffalo and Human Milk Proteins

Amino Acids	Human Milk (%)	Buffalo Milk (%)	Requiement of Protein level to Provide deficient amino acid (%)
Cystine	3.05	2.47	1.23
Histidine	2.27	2.48	Excess
Insolecine	5.01	6.04	Excess
Leucine	9.09	10.63	Excess
Lysine	6.74	6.49	1.04
Methionine	1.90	2.80	Excess
Phenylalanine	3.60	3.83	Excess
Threonine	4.31	3.08	1.40
Tryptophan	2.19	1.41	1.55
Tyrosine	2.82	3.73	Excess
Valine	6.50	8.80	Excess

Source : Joglekar, 1984.

WEANING FOODS

The importance of special protein for the weaned child has now been realized in many countries. While formulating the weaning food, two important considerations need to be kept in mind. One is the selection of raw material depending upon the availability and established tastes for particular ingredients and the other is the cost and resultant blend.

Ingredients

The choice of raw materials for weaning foods is wide, several cereals such as wheat, rice, maize, jowar in combination with various pulses/oilseeds can be used for weaning food formulations. Milk can also be used as one of the ingredients to improve the protein quality. These ingredients should be blended in such proportions as to minimize the deficiency of amino acids and the resultant blend to have a nutritive value superior to that of the ingredients from which it has been made. Vitamins and minerals are also added to weaning food.

Composition of Weaning Food available in Indian Market

The chemical composition of wheat flour based weaning food of India (Cerelac, Nestle) as printed on the label is given in Table 8.14.

Table 8.14

Composition of Wheat Flour Based Weaning Food of India (Cerelac, Nestle)

Attributes	Per 100 g of Product
Fat (g)	9
Linoleate (g)	1.5
Protein (g)	15.5
Carbohydrates (g)	68.9
Dietary Fibre (g)	1.4
Ash (g)	2.7
Moisture (g)	2.5
Energy value K cal/KJ	419/1750
Minerals	Per 100 g of product
Sodium (mg)	200
Potassium (mg)	470
Calcium (mg)	570
Phosphorus (mg)	380
Iron (mg)	7.5
Vitamins	
Vitamin A (IU)	1167
Retinol mcg	350
Vitamin D (IU)	200
Vit. E (IU)	3
Vit. C mg.	35
Thiamine (B_1) mg	0.8
Riboflavin (mg.)	0.6
Nicotinic Acid (mg.)	5
Vitamin B_6 (mg.)	0.3
Folic acid (mcg)	22.5
Panthothenic Acid (mg)	1.5
Vit. B_{12} (mcg)	0.75
Biotin (mcg)	25

Manufacture Technology : Three different types of cooking and drying processes are used for the manufacture of weaning foods.

(i) In this process precooked material is dried on the roller or spray drier.

(ii) In the second method, the blend is passed through a macaroni press, steamed and flaked and finally dried.

(iii) In the third method, the dry ingredients are extruded through an extrusion cooker.

Different Types of Weaning Food

(i) The spray dried weaning food developed from soy-dahl Skim milk powder and barley malt showed PER (2.7) as observed by Chandrasekara *et al.*, (1966).

(ii) Soy-whey weaning food developed in NDRI laboratory (Chandramohan, 1977) had a PER of 2.2 without fortification with methionine and 3.2 after fortification (as compared to 2.5 of Casein).

A spray dried soy-whey weaning food has been made (Fig. 8.3) from cheddar cheese whey (65 parts, on solid basis and blanched soybean cotyledons (35 parts).

(iii) PER of potato-Soybean-Skim milk weaning food (1.9) has been developed by Kaur (1980).

(iv) A jowar-Soybean-Skim milk (JSM) weaning food includes 60 parts jowar (sorghum) flour, 30 parts soybean solids and 10 parts Skim milk solids.

Fig. 8.3 : Flow diagram for the manufacture of Soy-Whey Weaning Food

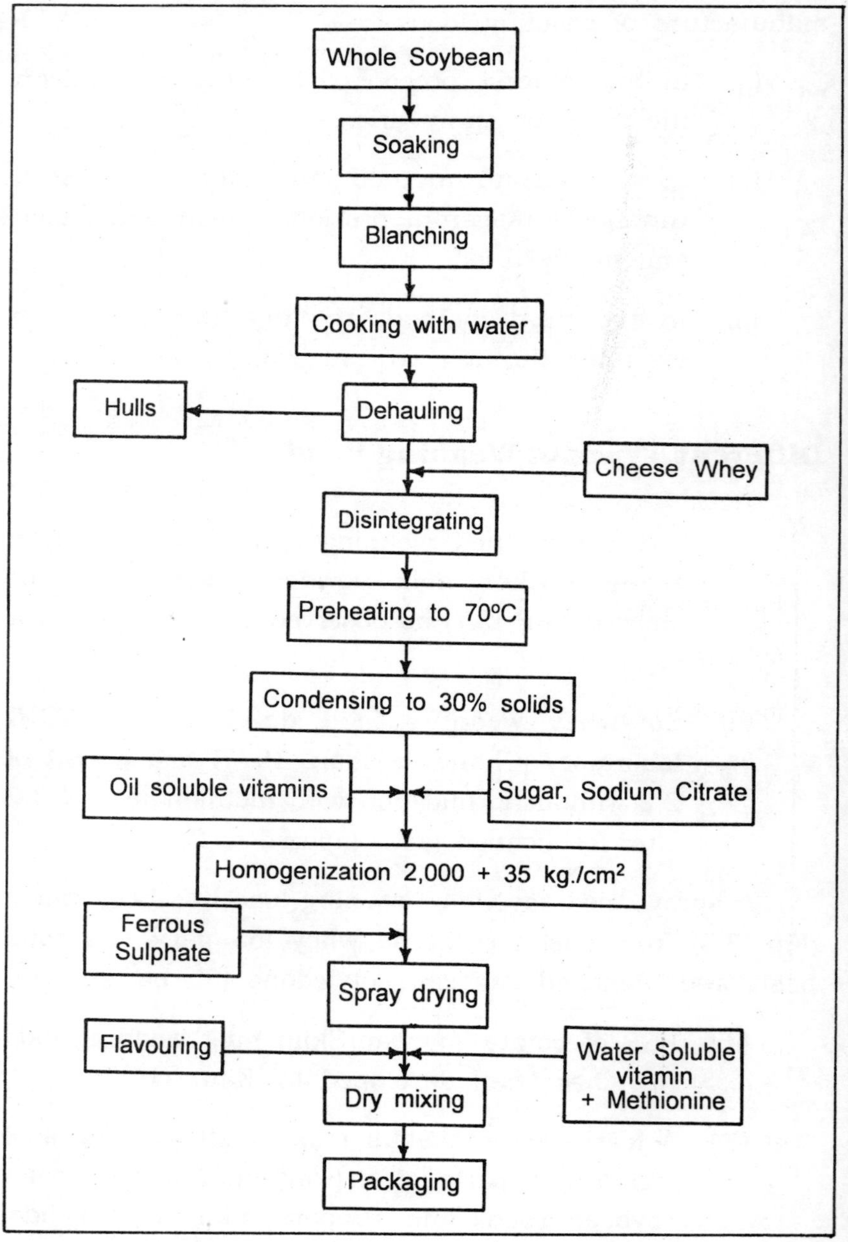

This has been specifically formulated to meet the need of jowar growing areas. PER of J-S-M weaning food (60 : 30 : 10) was (2.6). It is observed that 15% sugar level is most acceptable (Gupta and Arora, 1982). Raw material required for JSM weaning food is given Table 8.15 Proximate composition and protein efficiency ratio (PER) of Jowar-soybean–Skim milk weaning food (Table 8.16). Flow diagram for the manufacture of jowar-soybean-Skim milk weaning food is given in Fig. 8.4.

Table 8.15

Raw Material Required for JSM (60 : 30 : 10) Weaning Food

Raw Material	Quantity in kg.
(a) Jowar	75.0
(b) Soybean	42.0
Skim milk	123.5
Calcium Carbonate IP	0.3
Sodium bicarbonate IP	3.0
Sugar	15.0
Vitamin premix (B + C)	0.030
Vitamins A, D, E & K	0.088
Flavouring	0.092

Source : Gupta & Arora (1982)

Fig. 8.4 : Flow diagram for the manufacture of Jowar-Soyabean-Skim milk weaning food.

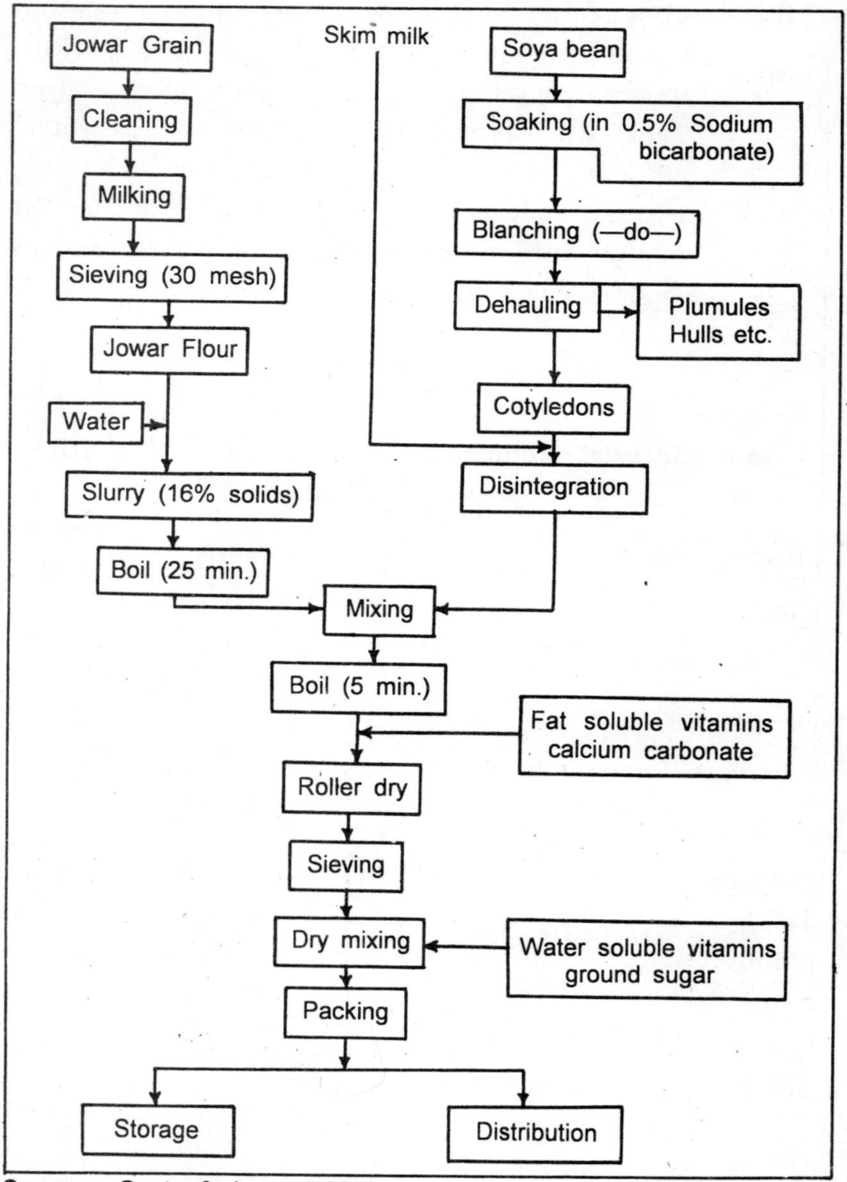

Source : Gupta & Arora (1982)

Table 8.16

Proximate Composition and Protein Efficiency Ratio (PER) of Jowar-Soyabean–Skim Milk Weaning Foods

Characteristics	J-S-M (60 : 30 : 10 %) Weaning food	Weaning food (PAG 1971)
Moisture, g%	7.5	5-10
Protein, g%	22.1	more than 20
Fat, g%	9.0	upto 10
Crude fibre, g%	1.7	less than 5
Total ash g%	3.0	less than 5
Carbohydrate g%	56.2	—
Calcium mg%	340.0	300 as Po_4 or Co_3
Phosphorus mg%	180.00	—
PER*	2.6	More than 2.1

* PER corrected to 2.5 of casein

Source : Gupta & Arora (1982).

MALTED MILK POWDER

An organisation known as Wisconsin Mellin Company manufactured first malt based food product in the year 1850. This product was branched as "Leigbig". It fed mainly to infants after mixing with sterile milk and water. Malted milk was the first dry milk product made commercially in the United States. It was invented in 1883 and appeared on the market in 1887. The manufacture of it was undertaken by William Horlick, at the request of physicians who wanted a baby food prepared from milk and cereals.

Table 8.17

PFA Standards of Malted Milk Powder

Characteristics	Requirements	
Moisture (max)	5.0%	
Total ash (max)	5.0%	dry matter basis
Ash insoluble in dil Hcl (max)	0.1%	—do—
Milk fat (min.)	7.0%	—do—
Nitrogen (min.)	2.0%	—do—
Solubility (min.)	80.0%	—do—
Starch test	-ve	
bacterial count/g (max.)	50,000	
Coliform count/g (max.)	10	

Table 8.18

IS : 1806 : 1975, Standards for Malted Milk Powder

Characteristic	Type I	Type II*
Moisture (max.)	4%	4%
Total protein (N x 6.25) (min.) dry	13	11.5
Fat (min.) dry	7.5%	7.6%
Total ash (max.) dry	5%	5%
Acid insoluble ash (min.) dry	0.1%	0.1%
Solubility (min.)	85.0	80.0
Cocoapowder (min.) dry	—	5.0
Test for starch	-ve	-ve
Bacterial count/g	50,000	50,000
Coliform count/g	10	10

* with cocoa powder.

Jones and William, two brothers from the U.K. started manufacturing an artificial food product by oven drying using an extract of malted barley and wheat flour. The product was filled in 14 ounce bottles and mixed with clean milk for making a drink. The product manufacturing process was modified with the addition of cow's milk to increase its nutritional value. Batch mashing, first and second stage evaporation and oven drying process was used for manufacturing first malt-milk based food product.

Definition

The US Federal Standard provides that dried malted milk powder shall contain not less than 7.5 per cent milk fat and not more than 3.5 per cent moisture and that each kg/lb of malted milk powder shall contain the total solids of approximately 2.2 kg/lb of whole milk.

Malted milk food is the product obtained by mixing whole milk or partially skimmed milk or milk powder with malt extract and cereal grain flour in such a manner as to secure complete hydrolysis of the starchy material. It should not contain added sugars, foreign fat and added colours and preservatives. It may contain Coca powder. Malted milk food should be in the form of powder.

Standard

Tables 8.17 and 8.18 show PFA and Indian Standards for malted milk food.

Composition

An average chemical composition is given in Table 8.19.

Table 8.19

The average composition (%) of the Malted Milk Powder

Moisture	Fat	Protein	Lactose	Ash
3.29	7.55	13.19	72.40	3.66

Vitamin and mineral composition of selected brand malted milk powder of India is given in Table 8.20.

Table 8.20

Vitamin and Mineral Composition of Selected Brand Malted Milk Powder* of India.

Characteristics	Bournvita	Boost
1	2	3
Vitamins		
Vit. A (mcg)	950	1000
Vit. C (mg)	70	75
Vit. B_1 (mg)	2.4	1.5
Vit. B_3 (mg)	33.5	—
Vit. B_6 (mg)	3.3	1.67
Biotin (mcg)	—	33.3
Vit. B_{12} (mcg)	1.8	3
Vit. D (mcg)	—	2.83
Vit. B_2 (mg)	—	1.5
Niacin (mg)	—	20
Patothenic acid (mg)	—	6.67
Folic acid (mcg)	375	500

1	2	3
Minerals		
Zinc (mg)	4.8	4.53
Manganese (mg)	1.8	—
Calcium (mg)	100.0	737
Iron (mg)	18.0	26.7
Magnesium (mg)	—	136
Copper (mg)	—	1.17
Energy (Kcal/100g)	382	396

*As indicated in label.

Nutritive value

Malted milk powder is a rich source of readily available carbohydrates. The cereal proteins originally present are partially digested during malting process. Malted milk is a good source of mineral salts, vitamin A and vitamin B complex. It furnishes about 382 Kcal per 100 gm.

Uses

Malted milk powder has a pleasant flavour and is used for feeding children, invalids and convalescents.

Ingredients Used in Malted Milk Powder

Milk solids : Increase nutritional value of the product by providing high quality fat, proteins, minerals and vitamins. Milk fat enhances flavour and energy level of final product.

Malted barley or malted wheat : These provide an appropriate level of â enzymes required to convert all the starch into simple sugars and also to impart a typical malty flavour to the finished product.

Wheat flour or any other source of starch : Wheat flour provides high level of simple sugars, proteins and fat in the finished product at low cost. All the starch is fully converted to simple sugars during the mashing process.

Malt extracts : A concentrated extract of malted barley provides desired level of colour and typical caramelized flavour to the finished product.

Salt : Salts are added to optimize product pH to improve digestibility, enhance flavour and to provide essential salts, specially calcium.

Minerals and vitamins : These micro ingredients would improve nutritional value of the product and provide essential requirement of minerals and vitamins.

Manufacture Technology

Testing of material : Malted barley, malt extract, wheat flour, milk powder and salt are either received in bags or in bulk containers. The raw materials are tested to conform to the laid down quality standards and stored either in bags or silos as per process design. The raw materials are then conveyed to various sections of the manufacturing process for usage.

Cereal extract preparation : Cereal extract preparation can be done by the following two processes.

(a) Batch mashing system : Malted barley is sieved to remove dust and other foreign matter. It is then crushed using in line roller mill and mixed with wheat flour. The mix is conveyed to mash tanks, where water at 40°C is continuously added to prepare an uniform slurry of ingredients. The slurry is continuously heated and stirred using a specially designed agitator. The slurry is gradually heated to raise the temperature from 40°C to 70°C within about 40 min. Slow heating is required to optimize activity of β type amylase enzyme for initial breaking of complex starch chains. This helps further conversion as shown in Fig. 8.5.

Mix Concentration

First stage concentration : The mix at about 26 per cent solids is concentrated under 680/690 mm. Hg. Vacuum, either in single stage calandria type or in multiple stage evaporators. The concentration is normally achieved upto average 58 per cent solids, depending upon the drying process used. Single stage evaporators were traditionally used in batch process to improve colour and flavour of final product.

Single stage evaporators have been replaced by multiple effect evaporators to save energy and labour cost. Multiple effect evaporators design has been modified to improve flavour during concentration.

Second stage concentration : Discharge from first stage is further concentrated upto an average of 80 per cent solids in specially designed second stage evaporators. The concentration is done under vacuum of 700 mm. Hg. using speci·lly designed agitator. The agitator keeps the

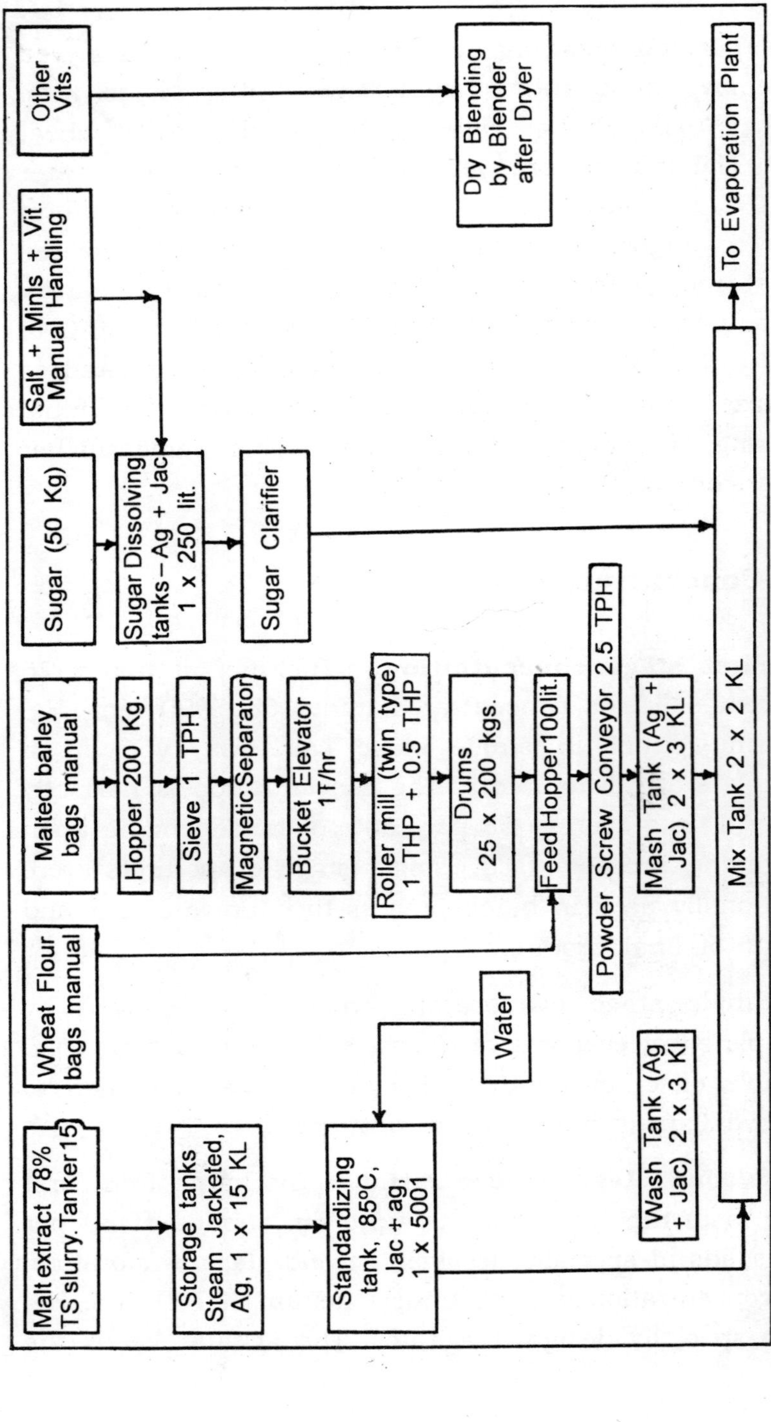

Fig. 8.5 : Flow Chart for Malt-mix Preparation

Source : Dhillon (2005)

thick slurry moving to avoid surface burning and improve water evaporation. The second stage evaporator is provided with steam heated jacket and drum to provide maximum heating surface. The evaporation tune varies from 2.5 to 3 hr. per batch depending upon the batch size and product formulation. The concentrate develops a typical malty flavour during concentration. The final concentration is discharged from the evaporator and pumped to steam jacketed storage tank for further processing.

Drying of concentrate : The concentrate can be dried by using Vacuum oven drying, band drying and spray drying. The typical flow diagram of the three drying method is given in Fig. 8.6.

Vacuum oven Drying : The concentrate of about 80 per cent from second stage is stored in a steam jacketed tank to keep it warm. The concentrate from the holding tank is pumped to the hopper of spreading machine. The concentrate is squeezed out through a slit from the hopper to spread a uniform layer of concentrate on the steel trays moving under the spreading machine. Trays are then moved into steam heated oven shelves. Saturated steam at about 1 kg. pressure is circulated through the shelves to provide necessary heat for efficient water evaporation. The concentrate is dried within a level of 2.0 per cent moisture under 710 mm. Hg vacuum. Dried food is of starch to simple sugars by the action of amylase enzyme.

The slurry is held at 70°C for about 35 min. for complete conversion of starch into simple sugars mainly maltose and dextrose. The slurry temperature is gradually raised to 78°C and held at this temperature for about 20 min. to stop further activity of enzymes and also to

Fig. 8.6 : Flow chart of drying Method for Malt Product

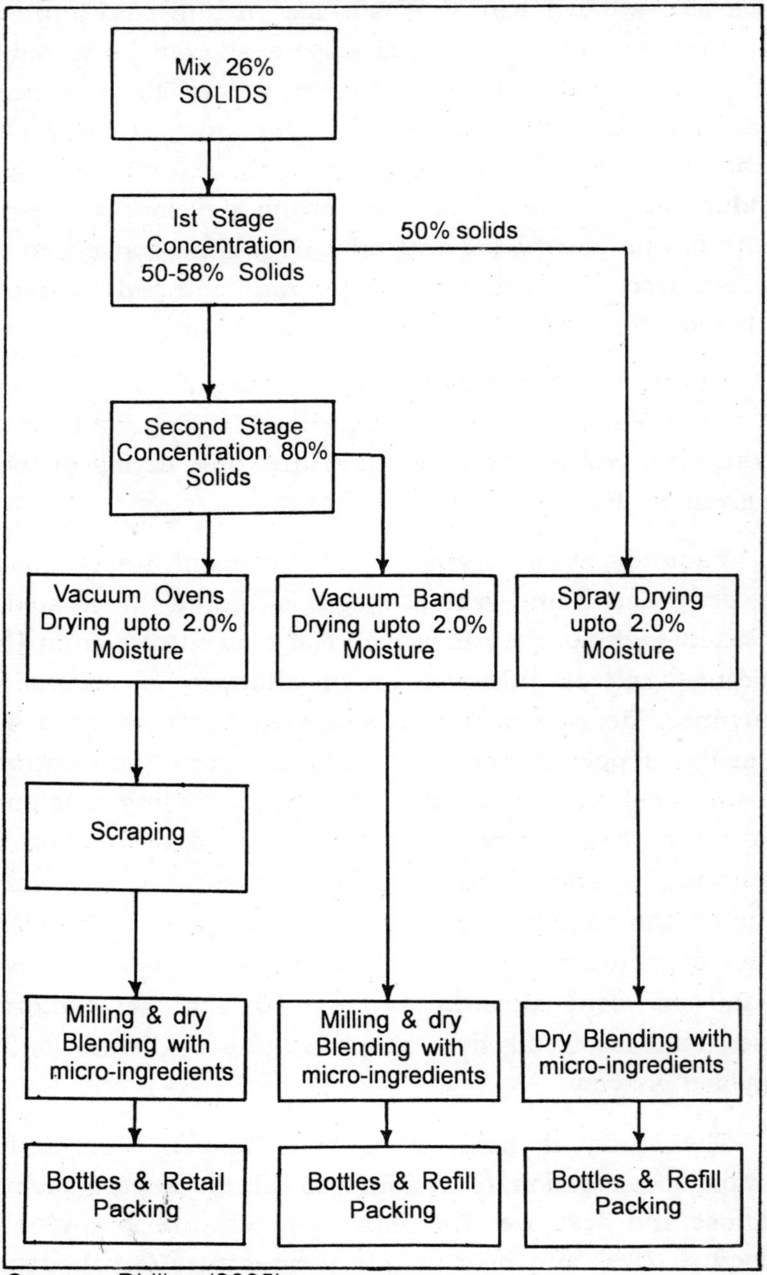

Source : Dhillon (2005)

achieve maximum cereal solids (28 to 30%). The slurry is then pumped to filtration system to separate barley husk and to obtain a clear liquor, known as 'wort'.

(b) Continuous mashing system : Rows of tubes arranged in series are immersed in hot water baths, maintained at different temperature. Complete starch conversion takes place as the slurry of wheat flour and malted barley travels through these tubes. The principle of slurry heating and time/temperature combination are similar to those in batch mashing system.

Advantages of continuous mashing over batch mashing system : The continuous mashing produces uniform quality wort at low energy and low labour costs, as compared with batch mashing system.

Malt extract standardization : Malt extract is received in bulk is pumped to standardization tank, where it is diluted with the addition of hot water and pasteurized by heating upto 85°C before adding to mix preparation tank.

Milk standardization : Pasteurized milk is standardized to the required fat/SNF ratio with or without addition of milk powders.

Handling of sugar, salts, vitamins and minerals : Sugar is sieved and dissolved in hot water. The mix of salts, minerals and vitamins is added to sugar solution. This mix is clarified and added to mix tank.

Mix preparation : The cereal extract from mashing selection known as wort is pumped to the mix preparation tank, using in line monitoring system to control wort solids. Wort from mashing section is mixed with predetermined quantities of milk, malt extract, sugar, salts minerals and vitamins. The percentage contribution to each ingredient depends upon product final formulation.

A typical flow diagram for mix preparation is given us cooled in an air conditioned room to avoid moisture pickup. The trays are then sent to scraping section.

Scraping, milling and packaging : The dried product is scraped off the trays using a semi-automatic scraping machine. The scraped product is normally in the form of cakes of various sizes. The scraped product is conveyed to the milling section, where it is milled into fine powder using inline milling machines. Milled product is conveyed to inline mixers to mix salts, dry vitamins and minerals. The final product is filled in bottles or refill packs.

Band drying : Band drying is advanced technology over Oven drying. The concentrate at 80 per cent solids is pumped from storage tank and uniformly spread on the moving fine wire mesh band. The fine wire mesh band with uniform layer of concentrate moves over bands of steam-heated chests. The drying takes place as the wire mesh band moves under vacuum through the entire length of the bond dryer (about 150 ft.). The wire mesh band moves at a very slow speed. It takes about 45 to 50 min. for the band to travel one length of the dryer. The concentrate is dried to about 2.0 per cent moisture level. Dried product is removed from the band dryer using vacuum lock system at the end of the dryer. The wire mesh band rotates and the drying cycle continues. The product is milled using in line hammer or miracle mills. Salts and other micro-ingredients are mixed in the milled product. The product is filled in bottle or refills.

Spray Drying : Spray drying of malted milk powder is a technological development to provide consistent product quality with better dispersibility and wettability. The concentrate from 1st stage evaporation at an average of 50 per cent solids is atomized either rotary or nozzle atomizing system. Malted milk powders are highly hygroscopic, due to high level of added or process converted sugars. Proper drying of these products is possible using modified design of three stage dryers, preferably having an air-broom system.

Sanyal and Balachandra (1989) studied on changes in physico chemical properties of spray dried malted milk produced from 30 to 40% TS mix and stored in polyethylene pouches and glass bottles at 30 ± 1°C for 2 months. They found that physico-chemical changes were greater in the case of samples packaged in polyethylene pouches compared with glass bottles. Malted milk made from 40% TS mixed showed better results that made from 30% TS mix.

Function of Malted Milked Powder :

1. Cater to consumers requirement as a milk substitute in milk deficient areas and to people who cannot digest pure milk.
2. Human nutritional value providing necessary level of minerals and vitamins.
3. Easy digestibility and assimilability in all types of individual consumers.
4. Value addition to milk based products.

COFFEE/TEA WHITENERS

Coffee whiteners are intended as an addition to coffee in place of cream. They can be based on milk, but some include only the milk compent casein in the form of sodium caseinate or may not contain protein at all. In those products, vegetable fat is substituted for milk fat and balance is ensured by stabilizing salts. The product also contains sugar (cereal syrup), additives for taste correction and emulsifiers. (Pallansch M.J. 1970; Tran & Einerson, 1989; Buchheim & Desmek, 1990).

Function : The primary function of a coffee or tea whitener is to impart the desired white colour to the beverage to which it is added. It also imparts a desirable flavour and body to the beverage.

Types of Coffee Whiteners : Coffee whiteners as sold in the western countries are in three forms; powder, liquid and frozen liquid. All these products are basically oil-in-water-type emulsions which are dried or carefully concentrated, and capable of forming an emulsion when added to an aqueous medium such as coffee or tea. The stability and functionality of these emulsions depend primarily upon the degree and type of emulsion.

Ingredients of Coffee Whiteners and Their Function

Vegetable fat : Provides whitening powder and consistency.

Protein : Helps in emulsification, provides whitening power and consistency. It improves flavour.

Sucrose : It provides sweetness, consistency. It reduces the freezing point of the emulsion.

Corn syrup solids : It provides greater çonsistency with less sweetness.

Emulsifier : It helps in emulsification, improves disperibility and wettability.

Colloid stabilizing agents : It improves colloidal stability, imparts consistency.

Stabilizing salts : It improves colloidal stability and reduced tendency to syneresis in the frozen product.

Flavour and colour : It produces flavour and colour to the product.

Ingredients of Coffee Whitener : A typical formulation for coffee whitener ingredients is given in Table 8.21. A liquid coffee whitener based on groundnut proteins besides Skim milk solids developed by NDDB is sold in Baroda for the past several years.

Table 8.21

A formulation for powdered coffee whitener

Ingredient	Concentration (%)
Vegetable fat	10.0
Protein	2.0
Sucrose	2.5
Corn Syrup solids	2.5
Emulsifier	0.5
Colloid Stabilizing agents	0.15
Water	100
Flavour	q.s.
Colour	q.s.

Source : Glicksman, 1975.

Technology of Manufacture of Milk-based Coffee Whitener

Milk-based coffee whiteners are manufactured in the following way. The blend of condensed skim milk, sugar and stabilizer is pasteurized, pre-heated at 74°C and dried. If the product contains milk fat, the standardization of fat-to-milk-nonfat-solid ratio and homogenization are performed prior to concentration. The next operation is blending. If there is fat in the blend, it must be an emulsifier as well, and is more intensively treated products without fat (e.g., at 82°C for 5 minutes). Coffee Whiteners are usually instantized to improve their solubility.

COFFEE COMPLETE AND TEA COMPLETE

These are two convenient formulations developed at NDRI. These ready-to-reconstitute products contain all ingredients except water required in the final product. The coffee complete, based on sweetened condensed milk and instant coffee is in a condensed thick liquid form. The tea complete is a spray dried tea extract containing milk solids and sugar. The composition of these product is given in Table 3.22.

Table 8.22

Composition of Coffee Complete and Tea Complete

Constituents	% by weight	
	Coffee complete	Tea complete
Milk fat	8.5	13.0
Non-fat milk solids	20.0	31.5
Sugar	39.0	39.5
Coffee solids	7.5	—
Tea Solids	—	14.0
Moisture	25.0	2.0

Source : Bhanumurti 1977.

High Protein, Lactose Free Milk Powder

A new method for the production of lactose free, low-fat, high protein, dried milk was developed (Mistry & Hassan, 1990). 180 kg Skim milk containing approx 3.2%

Table 8.23

Average Composition of High Protein Lactose Free Milk Powder

	A.V. composition of (%)				
	TS	Total protein	fat	lactose	ash
Skim milk	8.61	3.22	0.08	4.91	0.73
Diafiltered Skim milk	21.57	18.9	0.5	0.08	1.67
Diafiltered SMP	—	83.9%	2.27%	0.73	7.05 (Moisture 5.3)

Solubility in water in room temperature was not complete, but it improved at higher temperature.

protein was ultra filtered at 38°C to 15% protein and 3.8% lactose. The retentate (ultrafiltered milk) was diafiltered 3 times at 32°C with water to approx. 18% protein and 0.1% lactose and 21% TS. This diafiltered skim milk was spray-dried in a pilot plant. Niro Spray dryer equipped with a rotary atomizer and gas-fired heater. To minimize heat denaturation of whey proteins, minimum possible processing temperature was selected. Inlet and outlet air temperature, was 120-125°C and 75-80°C respectively. Skim milk, ultrafiltered milk, diafiltered milk, permeats and powder were analysed for TS, total protein, fat, lactose and ash (Table 8.23). Solubility in water in room temperature was not completed, but it improved at higher temperature.

DISCUSSION QUESTIONS

1. Define Baby Food. Explain its legal standard and nutritive value.
2. How will you manufacture Baby Food ?
3. Define Fortification. Why is it necessary for Baby Food ?
4. What do you understand by Weaning Food ? How will you manufacture it ?
5. Explain the process of manufacturing Malted Milk Powder.

9

Dried Milk Product

Last two decades witnessed exuberant growth in dried milk product sector. Dried milk products are diversified form of dried milk with common origin and technological set up. It also serve the same purpose of saving storage space, reduce packaging and transportations cost and improves the storage life of the product concerned. The production of these products has the advantage of processing and storing all market surpluses for longer duration. Low water content in dried milk products also exhibits the development of microorganism, thus, increases the much awaited keeping quality of the product. Different dried milk products are Anhydrous milk fat, butter powder, cheese powder, lactose powder, Khoa powder etc.

ANHYDROUS MILK FAT

Commercial production of AMF emanated during the second world war. It has long been used in equitorial climates because of its extended shelf-life, regardless of storage condition. It consist of more than 99% milk fat, which is acquired from either butter or cream. Various processing technology were used in New Zealand, Australia, Germany, Canada, Switzerland and USA.

Advantages of AMF

- Extended Shelf life
- Needs no refrigerated storage
- Flavour of butter fat is not altered
- Flexible for various uses.

Uses

AMF can be used for producing

- Recombined milk
- Chocolate, toffee in the confectionery industry.
- Evaporated and sweetened condensed milk.
- Baby food and ice-cream.
- Recombined butter production.

Legal Standards

Anhydrous milk fat quality is determined by the International Standard (68A, 1977) of the International Dairy Federation (IDF) according to its composition (Table 9.1).

Table 9.1

Composition of Anhydrous Milk Fat

Characteristics	% AMF
Milk fat (minimum)	99.8%
Water (maximum)	0.1%
Free fatty acids (maximum)	0.3% (expressed as oleic acid)
Cu (maxi.)	0.05 ppm
Fe (maxi.)	0.2 ppm.
Peroxide Value not higher than	0.2
Neutralizing agent	No Traces
Coliform Microorganisms (1 g)	No Traces
Taste & Smell at 20-25°C	Clear

Production Technology

The technological process for production anhydrous milk fat are of two types :

(i) Frozen butter is melted, then concentrated.

(ii) Cream emulsion is split with phase inversion.

I. Anhydrous Milk Fat Production from Butter

Frozen butter is gradually melted in continuous butter melter and subsequently heated in a plate heat exchanger then transferred to a holding tank. Tank (Fig. 9.1) serves to agglomerated proteins in the melted butter and to separate air, to faciliate further fat separation. In the separator, the water phase, i.e., 'butter milk' is separated, leaving concentrated milk fat. Since the fat in this phase still contains traces of water, it is necessary to dry it in a vacuum drier. The plate heat exchanger serves to preheat the concentrated milk fat and after drying, to cool it by heat regeneration.

If butter of increased acidity is used as raw material, prior neutralization with alkaline solution must take place. However, when salted butter is processed, it is first washed.

II. Anhydrous Milk Fat from Cream

In this process, the raw material is cream, with a fat content of 35-40%.

Cream is pasteurized in a plate heat exchanger, cooled to 55°C and subsequently concentrated in a centrifugal separator to 70-75% of its milk fat content. Buttermilk from the centrifuge is separated in a separator and the cream, thus, obtained enters the process through a float hopper and proceeds to the heat exchanger. It is cooled in the heat regenerating section of the pasteurizer and discharged from the process. Concentrated milk fat is introduced into a centrifixator where fat globule

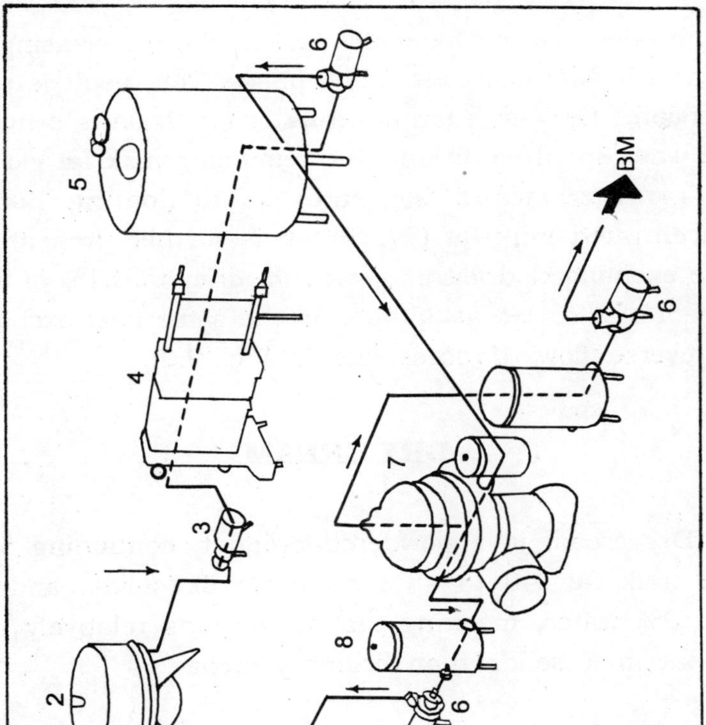

Fig. 9.1 Process line of anhydrous milk fat production (conversion of butter)

1. 2. Butter melter, 3. Positive displaement pump, 4, 9. Plate heat exchanger, 5. Cooling tank, 6. Pump, 7. Separator, 8. Float hopper. 10. Vacuum drier, 11. Storage ...

F = Feed (frozen butter), BM = Buttermilk, P = Product.

membranes are mechanically broken down, releasing free fat, which becomes continuous phase. The result is phase inversion : O/W–w/o (an oil–in-water emulsion is converted to a water-in-oil emulsion). The remaining milk fat globules are returned via a separator to a floating hopper. Concentrated milk fat (99.5% fat) is further preheated in plate exchanger, dried in a vacuum drier to 0.1% of water content and cooled at 35-40°C in the same heat exchanger by reverse flow. (Process line in Fig. 9.2).

DRY CREAM

Dry cream is a powdered product, containing 50 to 60% milk fat, 40 to 48% non-fat milk solids, and less than 2% water, meaning that it contains relatively more non-fat milk solids than ordinary cream.

Use

The use of dry cream is quite limited.

- Direct household use
- Used in confectionery, bakery products.
- Manufacture of foods where dry milk solids, primarily milk fat are required.
- For making Chocolate ice-cream.

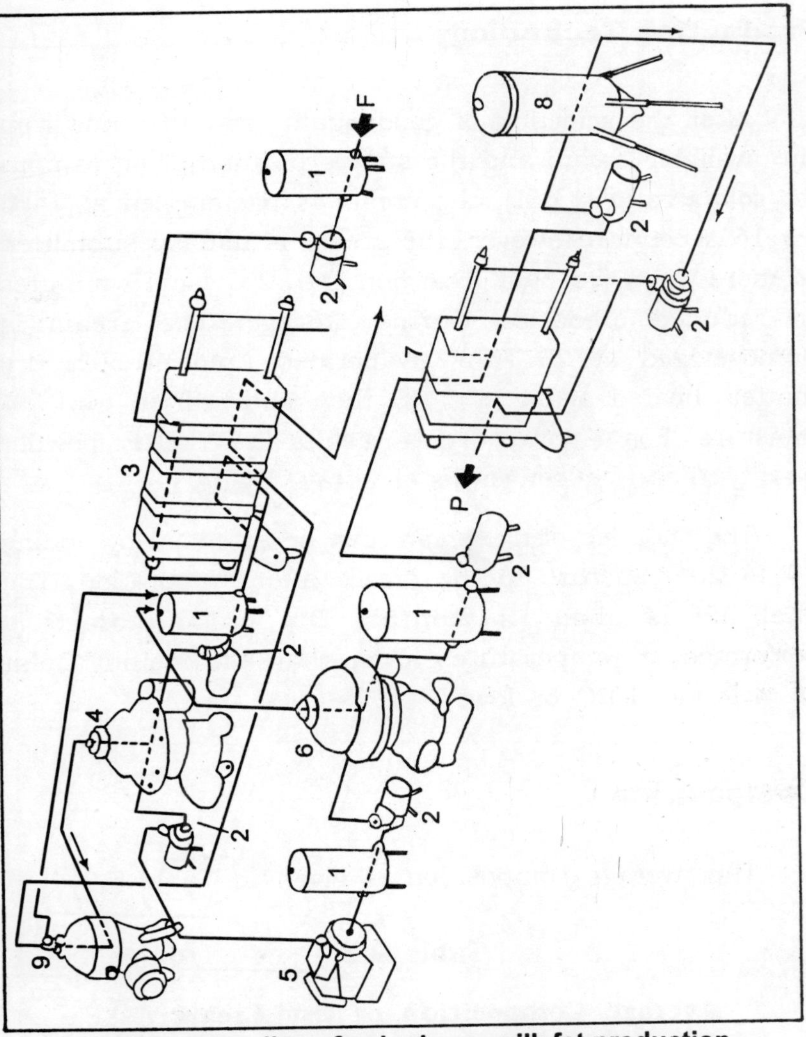

Fig. 9.2 Process line of anhydrous milk fat production (conversion of cream)

1. Float hopper, 2. Pump. 3, 7. Plate heat exchanger, 4. Centrifuge, 5. Centrifixator, 6. Concentrator, 8. Vacuum drier, 9. Separator.

F = Feed (cream with 35% fat content), BM = Buttermilk, P = Product.

Production Technology

After the selection of good quality raw material from the available cream and the standardization of fat to non-fat solids ratio (1 : 1), the cream is pasteurized at 74°C for 16 seconds. However, the cream is usually submitted to more intensive heat treatment at 82°C for 15 minutes to reduced oxidation during storage. The cream is homogenized at 20 MPa, evaporated upto 40-50% dry matter, heated again at 71°C then spray dried to 1-2% moisture. Foam spray drying results in a better powder quality than the conventional spray drying.

The shelf-life of dry cream can be extended by adding 10 to 20% sucrose to the cream after evaporation. The shelf-life is about 2 months. Dry cream-storage is performed at temperatures lower than the melting point of milk fat, 10°C or lower.

Composition

The average composition is given in Table 9.2.

Table 9.2

Average Composition of Dry Cream (%)

Content	Average %
Milk fat	65%
Protein	13.4%
Lactose	10.0%
Ash	2.9
Moisture	1.0

BUTTER POWDER

Composition

Butter powder is a rich source of butter fat. The average chemical composition of butter powder is given in Table 9.3.

Table 9.3

Average Composition of Butter Powder (%)

Content	Average %
Fat	81.98
SNF	6.70
Caseinate Citrate	6.70
Emulsifier	3.50
Free flowing agent	0.50
Anti-oxidant	0.02
Moisture	0.06

Uses : It is used for ice-cream making, bakery products in dry mixes, sauces, icings and fillings.

Production Technology

Ingredient

Table 9.4 shows the ingredient use.

Table 9.4

Ingredient Used in Butter Powder Making

Ingredient	(kg.)
Pasteurized cream (32% fat)	132.0
Dry acid casein	6.0
Sodium citrate	1.0
Skim milk (9% TS)	20.0
Sodium hydroxide (2%)	7.0
Glycerol mono-stearate (36%)	3.5
Sodium aluminium Silicate	0.5
Butylated hydroxy anisole	10.9

The production consists in mixing pasteurized cream and reconstituted skim milk powder to have 80% fat and 16.5% SNF (on dry basis). Sodium citrate is dissolved in the cream and homogenized at 140 + 35 kg/cm^2 at 65°C and pasteurized at 75°C/min, cooled and ripened to 0.16% lactic acid by using starter culture. Glyceryl mono stearete and BHA are blended and spray dried at 175°C inlet and 80°C outlet temperature. The product is cooled and dry blended with sodium aluminium silicate and 20%. Calcium phosphate as free flowing agent. The anticaking agent is also dispersed in water and applied to the walls of the dryer. The product is quickly cooled and removed from dryer.

Packaging and Storage

Normally, butter powder is packed in bags with polythene liners, tin cans etc. It may be stored at room temperature or in cooled room 10-15°C.

Shelf-life

The shelf-life is around 6 month at 10-15°C with gas-packing.

CHEESE POWDER

Often, hard cheese is converted into powdery form. Two major varieties of cheese powder are cheddar cheese powder and parmesan cheese powder. These powder possess immense nutritive value specially in terms of fat, proteins and minerals content.

Classification

Cheese powder may be classified according to the type and level of ingredients used in the formulation. These are of :

(i) Natural : Which made using natural cheese, emulsifying salts and optionally, natural cheese flavour.

(ii) Extended : Which incorporates other ingredients, such as dairy ingredients (e.g., skim milk solids, whey, lactose), starches, maltodextreins, flavours, flavour enhances and/or colours.

Alternatively, cheese powder can be classified according to the proportion of cheese solids, as a % of total dry matter (Missel, 1996).

(i) High cheese solids (i.e., ~95%, w/w)

(ii) Medium cheese solids (>50%, w/w)

(iii) Low cheese solids (<50% w/w);

Use of Cheese Powder

- Cheese powder is generally used as flavouring ingredient in a wide variety of food, especially snack coatings (e.g., popcorn, nachos, tortilla shells). extruded snacks, cheese sauces, soups, savoury dressing and savoury biscuits. (Missel 1996).
- The development of 'bake-stable' cheese powder is claimed to overcome problems encountered during high temperature processing (Lewin, 1996), e.g., unpredictable rise during the baking of crackers, off-flavour development during the production of extruded palated snacks.
- Cheese powder may also be combined with various spices, such as cumin, garlic, chilli or onion, for the production of seasonings or savoury additives.

Anonymous 1993, Lewin 1996, Hortenson, 1999)

Cheese Powder Manufacture Technology

The individual processing steps have been described (Hedrick, 1981, Guinee *et al.,* 1993, 1994, Missel, 1996, Darrington, 1999) and are discussed briefly below :

(a) Formulation of the Slurry Blend

The blend usually consists of commuted natural cheese, water, emulsifying salts, flavouring agents, flavour enhancers, colour, anti-oxidants, such as propy gallate, butylated hydrozyanisole (BHA) and/or filling material such as whey or skim milk solids, starches, maltodextrins and butter fat. In addition to antioxidants, fat encapsulation technology, which reduces the level of free fat in the powder, may be used to reduce the susceptibility to oxidative rancidity.

Generally, mature cheese, with an intense flavour is used so as to impart a strong flavour to the final product.

Apart from their lack of flavour-imparting properties, young cheese with a high level of intact casein are unsuitable as they result in very viscous slurries, which are difficult to atomize and dry efficiently.

Flavouring agents, such as EMC, hydrolysed butter-fat, starter distillates.

Flavour enhancers, e.g., Sodium chloride, monso-sodium glutamate, autolysed yeast extract.

Filling materials in extended cheese powder are usually added to replace cheese solids and thereby reduce the formulation costs. However, they may influence the flavour, wettability and mouth-coating characteristics of the product in which the cheese powder is used.

Typical formulations of the slurries requried for natural and extended cheese powder with different levels of cheese solids are given in Table 9.5

Table 9.5

Typical Formulations for Cheese Powder Low (25.5%, w/w, of total), Medium (54.3%, w/w, of total) and High (95.4%, w/w, % of total) Levels of Cheese Solids

Ingredient	Level of ingredients added during formulation (%, w/w, of total blend), prior to processing and drying		
	Low cheese solids powder	Medium cheese solids powder	High cheese solids powder
Mature Cheddar	—	17.0	32.0
Extra-mature Cheddar	20.0	19.0	33.0
Enzyme-modified cheese paste	1.0	0.2	—
Enzyme-modified cheese powder	—	2.0	—
Whey powder	12.0	5.0	—
Skim milk powder	8.0	3.8	—
Maltodextrin	16.5	11.0	—
Emulsifying salt	1.5	1.5	2.0
Sodium chloride	1.0	1.0	0.5
Butylated hydroxyanisole	0.2	0.4	0.5
Water and condensate	39.6	39.1	32.0

Based on data from Guinee *et al.* (1993a).

(b) Processing of the Blend and Slurry Formation

Processing involves heating the blend (using direct steam injection) to a temperature of 75-85°C in a Processed cheese type cooker or in large (e.g., 5000 L) dissolving tank with shearing blades, while continously shearing (e.g., at 1500-300 rpm). Maintaining the temperature < 85°C minimizes the loss of flavour volatiles and risk of browning, especially in formulations containing high levels of lactose or high dextrose equivalent malto dextrins. The blend is worked until the hot fluid slurry is homogeneous in colour and consistency (Fig. 9.3) incorporation of air should be avoided as it affects the physical (bulk density) and instant (wettability, dispersability) characteristics of the powder and its susceptibility to oxidation on storage (Masters, 1976, Tetra Pak Processing system, 1995).

(c) Homogenization of the Slurry

This step is optional but is commonly practised to ensure homogeneity of the slurry. The pressures applied (typically 15 and 5 MPa, first and 2nd stages, respectively) have a major influence on the viscosity of the slurry, with higher pressures generally imparting higher viscosity for a given level of dry matter.

(d) Spray Drying of the Slurry

- Commercially, cheese powder is normally manufactured using two stage drying systems, e.g., filternat (box) dryers are used frequently in the USA, while tall form dryers with one or more integrated fluidized beds are commonly used in Europe. These dryers influence the quality of the final cheese powder, the tall-form dryer is generally considered to give better flavour retention, larger powder particles and better powder flowability.

Fig. 9.3 Outline of production process for cheese powder

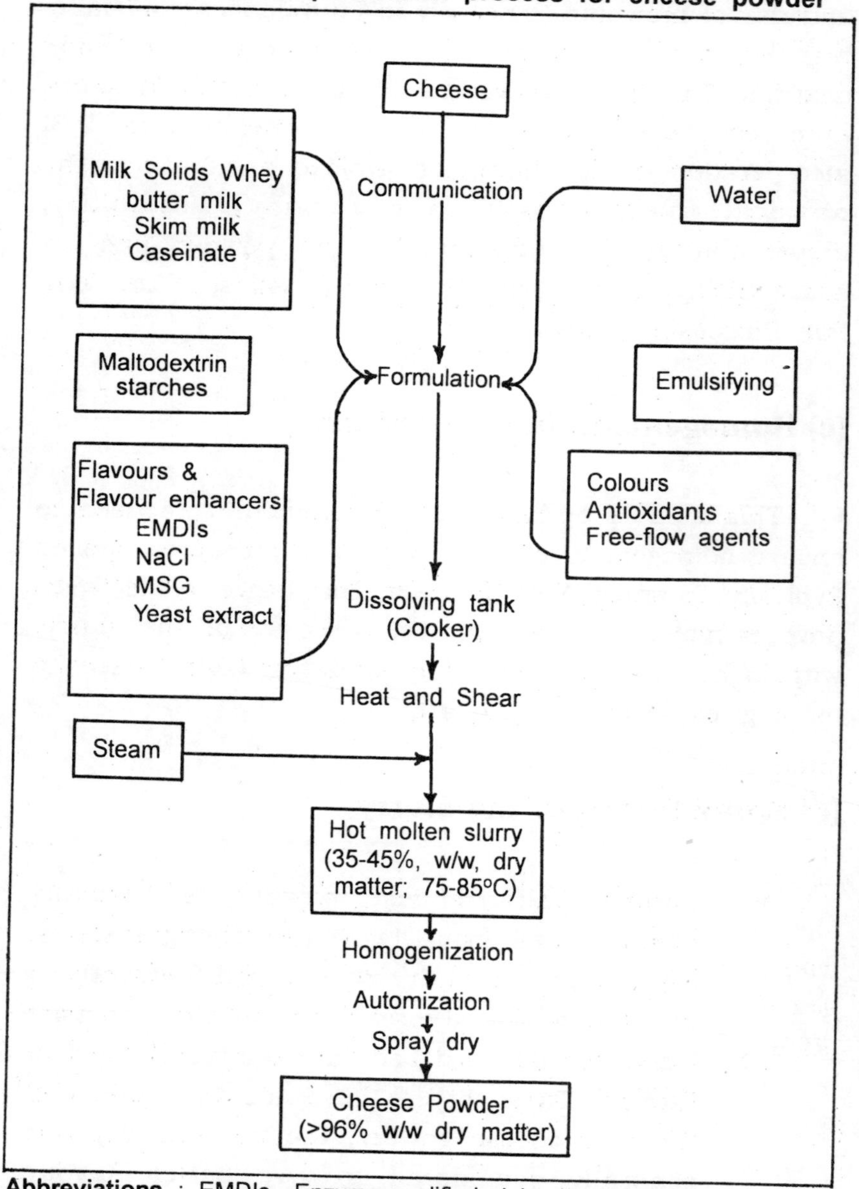

Abbreviations : EMDIs—Enzyme-modified dairy ingredients.
MSG—Monosodium glutamate.

Source : Guinee and Kiloawley (2004).

- Conventional, single-stage dryers are rarely used, except for experimental purpose. However single-stage silo dryers may be used, as in the Birs Dehydration Process (Kosikowski and Mistry, 1997).
- Typical air inlet temperature ranges from 180-200°C and air outlet temperature from 85° to 90°C depending on the dryer type.
- The powder is cooled to < 20°C in an external fluidized bed, using dehumidified air before packaging.
- The moisture content of the dried powder is typically 3.0-4.0%.

Pasteurized Cheese Powder

In 1984, an American patent was granted for the production of pasteurized cheese powder of the mozzarella type. In this process, milk is inoculated until pH is 5-9. The milk is subsequently cooled, ultrafiltered to a concentration degree of 3 : 1 whereby the pH is further decreased to 5.2. Through hydroxide addition, the pH value is adjusted to 6.3; the retentate is pasteurized by steam injection and again corrected. The retentate is finally dried. A pasty consistency is obtained through reconstitution, making the product suitable for pizza production.

DRIED CHEESE

The production involves fine grinding of hard cheeses and conveyance of the ground cheese to a dryed (usually fluidized bed-type) where it is exposed to low humidity air (15-20% kH) at an air inlet temperature < 30°C. Under these conditions, the vibrated cheese is dehydrated rapidly and evaporatively cooled, thereby reducing the risk of fat exudation, the tending to balling/clumping and borrowing. The dried grated cheese (typically 17% moisture) is generally pulverized and packaged under nitrogen to reduce the risk of oxidative rancidity during distribution and retailing.

High moisture (82%) cheese such as cottage cheese may also be dried directly to 3-4% moisture, by first pulverizing and then subjecting them to specialized spray-drying operations. (e.g., Silo spray drying using the Birs Dehydration process; kosikowski and Mistry, 1997).

Use of Dried Cheese

- Dried grated cheese is normally used as highly flavoured sprinklings (e.g., for pasta dishes) and in bakery products e.g., biscuits.
- Low moisture dried natural cheese is generally used for nutritional supplementation of foods, e.g., dried body meals.

LOW LACTOSE POWDER

Lactose intolerance is a global problem. There are people with total lack of lactose activity or having very small amounts of this enzyme. Ultrafiltration technology can be employed for the manufacture of low-lactose powder. Additional diafiltration treatment is employed to further reduce lactose. During the ultrafiltration process, some of the soluble salts like calcium, sodium and potassium are bound to go in the permeation. These salts are important for giving milk its natural taste. To maintain the salt level and thereby revive the original taste of milk on reconstitution, salts are added to the retentate before spray-drying. Further, to improve the drying properties of the retentate and reconstitutability of the powder 4% Malto-dextrin is added to the retentate before spray-drying. The benefit of this technology is that no chemical is added to reduce the lactose content in the powder with negligible loss of biologically available lysine.

Shelf-life

The product can be kept for about 9 months at room temperature.

DRIED ICE-CREAM MIX

Dried Ice-cream mix or ice cream mix powder became popular during World War II and later. There is considerable demand for ice-cream in summer season when milk production is low. That time, it is used for reconstitution into ice cream. It has become popular in India in recent years.

Composition

The average chemical composition of dried ice cream mix is given in Table 9.6.

Table 9.6

Composition of Ice-cream Mix Powder (%).

Moisture	fat	Solids Serum	Sugar	Stabilisers
1.0-2.5	25-29	25-30	37-42	0.5-1.2

Production Technology

The ingredients used are milk/skim milk/cream/skim milk powder/pregelatinized potato starch as stabilizer and tween 80 as emulsifier. The composition of the mix is adjusted so that it contains only 25% of the total sugar before drying. All the ingredients are weighed so that the mix contains 17.1% fat, 14.2% SNF, 5.3% sugar, 6.7% pre-gelatinized starch and 0.07% tween 80. The starch is mixed with a little amount of sugar and added to milk or skim milk and heated with agitation. The required amount of cream is now added and temperature raised to 50°C. Skim milk powder mixed with the remaining sugar is added to the mixture. The content is filtered and tween 80 added to the mixture. The temperature is raised to 70°C and homogenized at 175 kg/cm^2 and 35 kg/cm^2 in two stages. The mix is then pasteurized at 68°C/30 min., cooled to 5°C and aged for 16 h. The mixture is preheated and spray-dried with inlet temperature 73°C and atomizer speed 16,500 rpm. The powder obtained is dry blended with refined sugar in the ratio of 100 : 41.

Packaging

The product is packed in laminated polythene bags and can have shelf-life of 9 months at 30°C.

Reconstitution

For preparation of ice-cream, one part of the powder should be mixed with 1.6 parts of potable water at 45°C, cooled and frozen in freezer to about 100% overrun. A kg. of powder reconstituted in 1.6 kg. water may give 36 cups (125 mt.) of ice-cream.

KHOA POWDER

Khoa is a famous North Indian milk product, which is commonly used as a major base for sweet manufacturing. Due to its low self-life, it is transferred to khoa powder. Since it contains very low moisture (around 3%) its shelf-life remarkably increased than that of khoa. It is a highly nutritive product specially rich in fat, protein and lactose content.

Physico-chemical and Sensory Characteristics

Physico-chemical and sensory characteristics studied by different scientists are given in Table 9.7

Table 9.7

Physico-chemical and Sensory Characteristics of Khoa Powder

Characteristics	Patel & De (1977)	Thompkinson & De (1981)	Rizvi *et al.* (1987)	
	Buffalo	Cow	Buffalo	Cow
Moisture (%)	2.9	3.0	2.9	2.5
Fat (%)	33.9	31.3	33.9	29.0
Protein (%)	27.0	31.3	33.9	29.0
Lactose (%)	30.9	26.8	30.9	38.0
Ash (%)	5.3	4.8	5.3	4.0
Solubility index	—	13.0	14.5	—
(gl/ml)				
loose	0.26	0.33	0.27	—
Packed	0.35	0.43	0.35	—
Free fat (as % of total fat)	64.5	58.7	64.2	—
Colour	light brownish white	creamy yellow	white greenish tinge	—
Flavour	Pronounced Cooked Sweetish taste	Milk khoa like	Slight cooked	

Use : On reconstitution with water, this can be utilized directly for the preparation of burfi, milk-cake, kalakand and gulabjamun.

Production Technology

Various process tried for making khoa powder include Tray drying, Roller drying, Spray drying, fluidized bed drying and reverse osmosis.

(a) Tray Drying

the method of production of tray dried Khoa consisted of preparing it from standardized buffalo milk (5% fat and 9% SNF), heating the product at a suitable temperature to reduce moisture to about 20 per cent and comminuting it to fine particles. These particles are uniformly distributed in trays in the thickness of 1 cm. and dried in a cabinet dryer at around 70°C or alternatively in a vacuum oven (25˝ of Hg) (Fig. 9.4).

(b) Roller Drying

Standardized buffalo milk was first heated to a pre-determined temperature in a stainless steel double-jacketed vessel to develop a cooked flavour. The heated milk was pumped to the milk distribution channel of roller drier at a predetermined flow rate and steam pressure of 65 ± 5 psi. Khoa powder formed in thin sweets was scraped, ground sieved and packaged.

(c) Spray Drying

The manufacture of Khoa powder by the spray drying technique was standardized for large-scale operations. Milk was vacuum evaporated to 35% TS, heated to develop cook flavour and later, spray-dried at an inlet air temperature of 190°C and outlet air temperature of 88°C.

Fig. 9.4 : Khoa powder production from buffalo milk by tray drying (atmospheric pressure/vacuum)

Buffalo Milk

↓

Clarification

↓

Standardization
Fat 5%, SNF 9%

↓

Partial desiccation of Khoa to 20% Moisture (Steam 1.0 kg/cm^2)

↓

Breaking into bits

↓

Cooling at 30°C for 1 hour

↓

Pulverize to fine particles

↓

Cooling to 30°C

↓

Tray loading with Khoa particles 1.0 cm bed thickness

↓

Drying at 70°C

↓

Cooling

↓

Packaging

Drying is done for 4 hours under atmospheric pressure and for 3 hours under vacuum (25" of Hg).

(d) Fluidized Bed Drying

The application of fluidized bed drier for dehydration of khoa was also attempted. Finely ground khoa with 20% moisture was transferred to a fluidizing chamber and dried at an inlet air temperature of 92 + 1°C and outlet air temperature of 80°C.

(e) Reverse Osmosis (RO)

The feasibility of using RO concentrated milk for making khoa powder has been demonstrated. In this process, skim milk was first concentrated to 28% TS. The concentrate was then standardized by mixing 20% cream to obtain 6% fat in the final mix. The standardized concentrate was subsequently passed through a scraped surface heat exchanger to heat it to 126-130°C to impart the typical khoa colour and flavour. The heat-treated concentrated was then cooled down to about 65.6%°C and finally dried over a steam-heated roller drier to obtain khoa powder.

Comparison of Chemical Composition of Dried Khoa Prepared by Different Method of Drying

Table 9.8 shows chemical composition of dried khoa.

Table 9.8

Comparison of Chemical Composition of Khoa Powder Prepared by Different Drying Method

Characteristics (%)	Tray drying	Roller drying	Spray drying
Moisture	3.8	3.5	2.5
Fat	30.8	32.2	33.2
Protein	25.7	25.7	25.5
Lactose	34.7	33.0	33.9
Ash	5.0	5.0	4.9

Shelf-life

Khoa powder from tray, roller and spray processes could be preserved in tin containers at 30°C for five, six and seven months, respectively. Storage of product at 5°C extended the shelf-life to seven, eight and nine months, respectively.

GULABJAMUN MIX POWDER

The traditional method of Gulabjamun has several limitation such as non-availability of khoa all the year round, variation in the quality and that of resultant sweet and poor shelf-life. The formation of ready to use Gulabjamun mix powder eliminates all limitation found in Gulabjamun. In addition, it has many other advantages.

Advantages

- Does not require refrigeration storage.
- Produces sweets of very consistent quality throughout the year at relatively low cost.
- Consumers now prefer instant convenient food. Hence, it has very good market.
- It has very strong export potential.

Composition

Chemical composition of the gulabjamun mix is given in table 9.9

Table 9.9

Composition of Gulabjamun Mix

Characteristics	Percentage
Moisture	8.44
Fat	15.03
Total protein	19.45
Total ash	3.89
Carbohydrate (by difference)	53.19

Formulation

Formulation of gulabjamun mix is given in Table 9.10.

Table 9.10

Formulation of Gulabjamun Mix

Ingredient (%)	Skim milk powder based	Whole milk powder based
Roller dried (a) Skim milk powder	43.4	—
(b) Whole milk powder	—	58.4
Maida	25	35
Semolina (suji)	15	15
Butter fat/Vanaspati	15	—
Baking powder	1.5	1.5
Cardamom powder	0.1	0.1

In case of roller dried skim milk based formulation, all the dry ingredients are mixed thoroughly preferably in a dry blender. Fat, after melting, is incorporated slowly to ensure thorough mixing. The mix is packaged into printed polyethylene bags or laminated pouches and stored in a dry place.

Gulabjamun Preparation from Dry Mix

To make gulabjamuns, 100 grams of the mix powder is mixed with 50-55 ml of water or milk and is kneaded until the dough is of smooth consistency. Balls of spherical shape and uniform weight (10 g each) are rolled manually. Then, these are fried and soaked in sugar syrups.

Shelf-life

The shelf-life of dry mix powder in ordinary polyethylene bags at room temperature is more than 6 months. Gulabjamun mix powder packaged in metalized polyester PE laminate, without any preservative remains fit for use upto eight months at 30°C.

CHHANA POWDER

Chhana has a very very low shelf-life (1-2 days at room temperature). Storage of chhana has many problems. The production of dried chhana (low moisture content) is expected to have a much longer shelf-life than the original product.

Uses

For reconstitution into chhana for the preparation of indigenous sweetmeats. etc.

Composition

Composition of Chhana powder is given in table 9.11

Table 9.11

Composition of Chhana Powder (cow)

Moisture	Fat	Protein	Lactose	Ash
3.5	41.8	46.3	4.2	4.4

Production Technology

Fresh cow milk, standardized to 4.0% fat and filtered is heated to first boil, preferably in a stainless jacketed steam heated kettle of the rotary type, with constant stirring-cum-seraping to prevent the milk from scorching. The steam intake is then stopped. Sodium citrate at 0.02% of milk is now added as an aqueous solution and thoroughly mixed into the milk, which is then covered and left undisturbed at 80 ± 2°C for 2 ± 1/2 hours.

The milk is coagulated at about 70°C by the rapid addition of a 1-2% lactic or citric acid solution till the whey is clear. The mixture of coagulated mass and whey is strained through a bleached muslin cloth and the milk solids (chhana) obtained are cooled to room temperature by submerging the chhana—tied with a cloth without being pressed, in the form of a bundle—in running cold water.

The chhana is deposited in a separate vessel and broken up into small particles, either manually or with a suitable mechanical device. Then, at the rate of 19-21% water of total solid in the mixture is added. It is pre-heated to about 60°C and passed through a micro-pulverizer so as to give a smooth slurry. This is pre-heated to 80 + 5°C before spray drying using an inlet air temperature of 195 + 5°C and outlet air temperature of 102 + 2°C. The chhana powder is promptly removed from the drying chamber, cooled packaged and stored at room temperature.

Shelf-life

Chhana powder appears to have a marketable life at room temperature of 2 & 4 months under air tight and gas packed conditions, respectively.

DRIED ROSOGOLLA MIX

Rosogolla mix powder gives 20% higher yield than that obtained from traditional method. Rosogolla prepared with dried mix is reported to have blananced flavour and spongy texture.

Production Technology

A process has been developed for production of dried rosogolla mix. Skim milk is concentrated to a required total solids level by using ultra and dia-filtration process. A calculated amount of fat is added to the retentate before spray drying it. Certain additives and binders are dry blended in the retentate powder to yeild dried rosogolla mix.

Shelf-life

Dried rosogolla mix can be stored without spoilage for about five months at 80°C and for 10 months at 5°C in sealed containers.

INSTANT KULFI-MIX POWDER

It is formulated from milk fat, milk solids-not-fat, sucrose and isabgol husk. This product has a shelf-life of 7 months at 30°C in tin cans. The shelf-life can be extended upto 10 months with the additive of BHA and nitrogen gas flusing.

CHAKKA POWDER

Production Technology

Manufacture of chakka powder involves preparation of dahi followed by partial removal of whey from it. The chakka is then ground in a colloid mill to obtain a smooth and uniform consistency in slurry. The slurry containing fat is heated to 38°C and then homogenized in a single stage homogenizer at a pressure of 100 kg/cm^2. A spray drier is used to make the powder using 185°C inlet and 85°C outlet air temperature (Fig. 9.5).

SHRIKHAND POWDER

Production Technology

At first, the shrikhand base (i.e., chakka) is prepared under standardized condition, then good grade ground sugar is added to the base at the rate of 18% and thoroughly mixed into it by kneading. A calculated amount of water at 20-25% of the mix is now added to it and

Fig. 9.5 : Flow chart for the manufacture of standardized milk chakka powder.

Standardized milk (4% fat)

↓

Heat treatment (90ºC / 15 min.)

↓

Cooling to 22ºC

↓

Inoculation – lactic fermentation with LF – 40 @ 1%

↓

Incubation for 12-up hours

↓

Dahi

↓

Drainage of Whey

↓

Chakka

↓

Grinding in Colloid Mill

↓

Addition of Water

↓

Homogenization 100 kg/cm^2

↓

Dryer

↓

Chakka powder

well stirred to give a smooth slurry. The slurry is homogenized at 100 kg/sq. cm. pressure and spray dried under standard techniques using an inlet air temperature of 180-200°C and an outlet temperature of 100°C. The powder is promptly removed from the drying chamber, cooled, packaged and stored in room temperature with or without gas packaging.

Composition

Composition of chakka and shrikhand powder on dry matter basis is given in Table 9.12.

Table 9.12

Composition of Chakka and Shrikhand Powder on Dry Matter Basis

Characteristics	Standard Milk Chakka Powder	Standard milk Shrikhand Powder
TS	96.18	96.08
Fat	36.48	19.82
Protein	44.96	24.12
Reducing Sugar	11.53	6.64
Sucrose	—	43.44
Ash	3.21	2.06
Acidity (% 1A)	6.58	4.94

Packaging and Storage

Shrikhand powder may be packaged in paper bags with polythene liner or fibre drums with or without polythene liners, with or without gas packing for storage at room temperature.

Shelf-life

Dried Shrikhand seems to have a marketing life over 3 months at room temperature storage, when gas packed in hermetic containers.

KADHI POWDER

Composition : The average composition of Kadhi powder is given in Table 9.13.

Table 9.13

Average Composition of Kadhi Powder

Constituents	Percentage
Moisture	7-9
Fat	11-15
Protein	25-32
Ash	7-9
Total solids	90-93

Production Technology

For Kadhi preparation, dahi is made from milk standardized to 0.8-1.0% fat, Milk is pasteurized cooled to 37°C and inoculated with a mixed culture of *Lactococcus cremoris, Lactococcus lactis*, and *Streptococcus thermophilus.* This mixture is incubated at 37°C until an acidity of 0.95% (expressed as lactic acid) is attained. Addition of 4.5 to 5% besan (Bengal gram flour) with equal quantity of water is added to the stirred dahi. After mixing thoroughly, the mixture is heated to boiling and held at that temperature for 10-15 minutes. The slurry is cooled to about 60°C and dried on a double drum roller dryer. The recommended speed of roller is 16-20 rpm and a steam pressure of 40-50 psi (Fig. 9.6). Addition of various

Fig. 9.6 Flow chart for preparation of dry kadhi powder

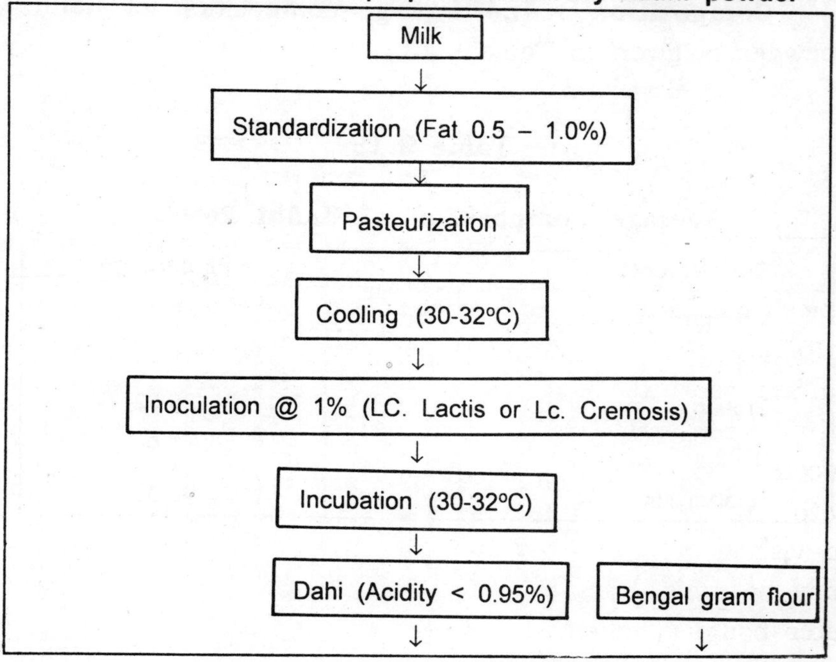

(Contd.)

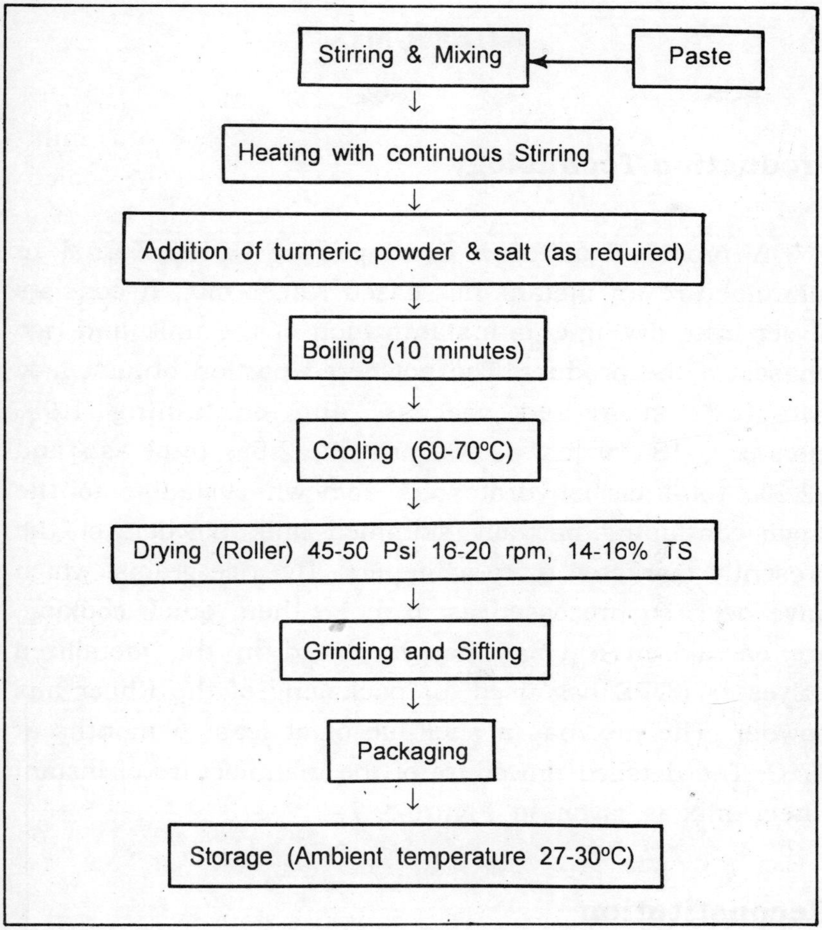

spices like dried chilli powder, mustard or cumin seeds into slurry before drying should be avoided because it causes serious colour defects. These should be added when reconstituting the dehydrate kadhi. For reconstitution, 1 part of kadhi powder is to be mixed with 4 parts of boiled water. The ratio can vary according to viscosity preference. Cut vegetables and curry leaves may be added before boiling fried besan balls (pakoras) after boiling (Ghosh and Kulkarni, 1990).

KHEER MIX

Production Technology

A process has been developed at NDRI, Karnal to manufacture an instant rice-based Kheer mix. It consists of separate drying-cum-instantization of the milk and rice phases of the product. The powdered portion obtained by two-stage spray-bed process, and containing 1.9% moisture, 18.2% fat, 15.3% protein, 2.5% total ash and 62.1% total carbohydrates, is somewhat similar to the sugar-containing partially skimmed milk powder, or the presently marketed dairy whiteners. The rice grains, which have been so processed as to make them quick cooking, are carried in a small pouch placed in the metallized polyester/LDPE bag used for packaging of the Kheer mix powder. The mix has a shelf-life of at least 6 months at 37°C. The detailed procedure of the manufacture of instant Kheer mix is given in Figure 9.7.

Reconstitution

Reconstitution of Kheer mix involves re-hydration of instant quick-cooking rice in boiling water for 10 minutes followed by dispersal of the powdered component into the rice-water mixture. The reconstituted product could be suitably flavoured and enriched with dry fruits etc. This product has been found to have a high acceptance rating in a consumer study (Jha, 2000).

Fig. 9.7 Flow chart for the manufacture of instant Kheer mix

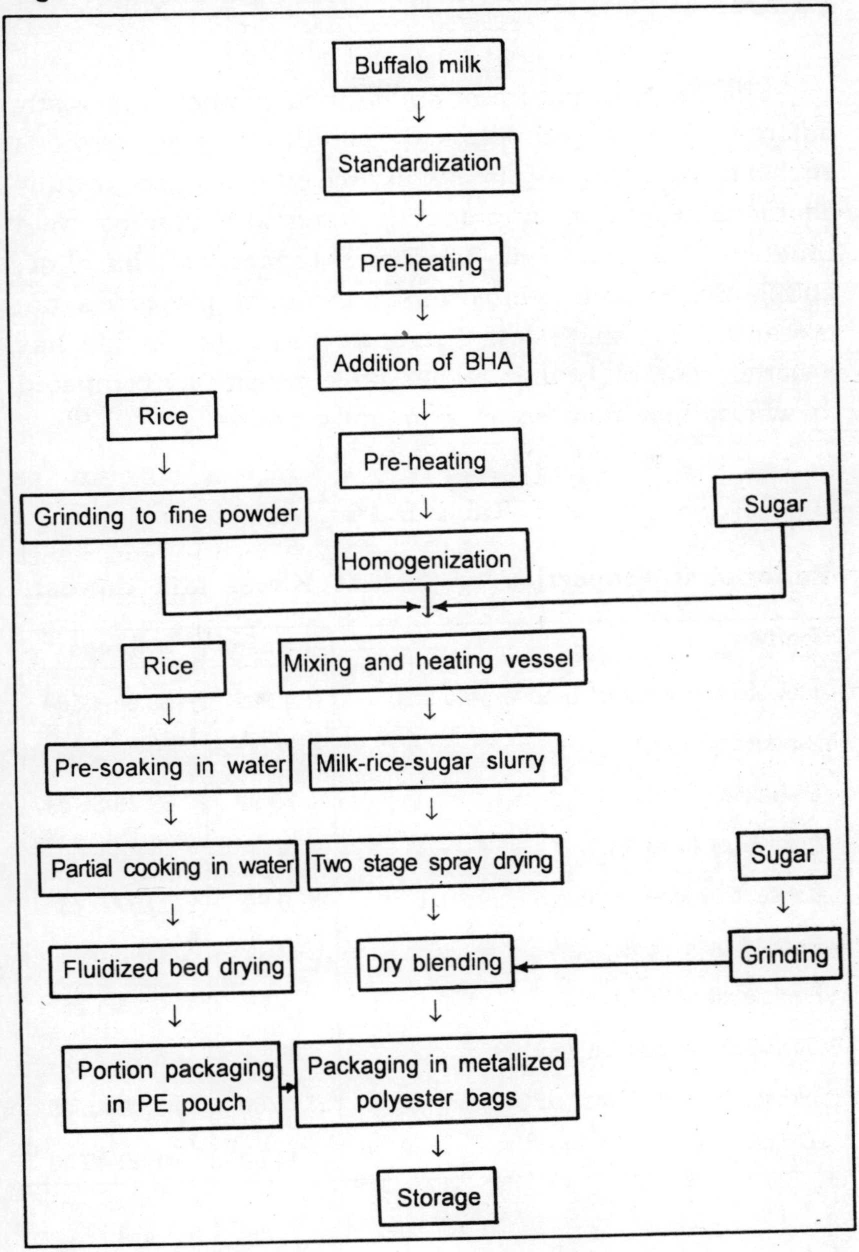

Functional Properties of Instant Kheer Mix Powder

The physical nature of spray-dried powders are vastly different from those obtained by other drying methods such as drum drying, freeze drying etc. In order to fully characterise the new product, systematic studies were conducted to analyse its functional properties (Jha *et al.*, 2002). Some of the important functional properties are presented in Table 9.14. The powdered milk fraction has a better flowability and higher bulk density as compared to whole milk powder or skim milk powder.

Table 9.14

Functional Properties of Instant Kheer Mix Powder

Property	Mean	Range
Flowability (Angle of repose, degrees)	40.09	38.30-42.64
Wettability (min.)	2.00	2.00-2.00
Dispersibility (%)	75.38	69.73-86.51
Insolubility index (ml)	4.00	3.00-5.00
Loose bulk density (g/cc)	0.69	0.67-0.73
Packed bulk density (g/cc)	0.81	0.80-0.83
Particle density (g/cc)	1.25	1.25-1.25
Occluded air content (cc/100 g)	6.63	5.90-7.00
Interstitial air content (cc/100 g)	45.00	45.00-45.00
Porosity (%)	44.80	41.60-46.40

DRY PALADA PAYASAM MIX

The development of palada Payasam dry mix with a long shelf-life has improved the commercial viability of the product in addition to helping the consumers to prepare the Payasam in a convenient way.

Production Technology of Dry Palada Mix

Five hundred grams of the dry ada flakes are added to 5 litres of boiling water and allowed to stand for 1 hour with occasional stirring, for the ada flakes to become soft. Later, the water is decanted and the ada is washed 2-3 times with cold water and the water drained off. The washed ada flakes are, then, boiled with 2 litres milk (fat 3%, SNF 8.5%) and 700 g of sugar in a steam-jacketed vat till it becomes a pasty mass. At that stage, sugar syrup made by boiling 1 kg. of sugar with half its quantity of water is added and the heating continued. Utmost care has to be taken at this stage to see that the contents did not stick to the sides of the vessel, as otherwise charring would take place. The sugar syrup helped to give a coating to the ada flakes, which would protect them from deterioration during storage. The heating is continued with constant stirring till almost all the water is driven out. Two hundred grams of powdered sugar are added at this stage as a seed material for the sugar to crystallise out. Otherwise, the content would remain as big lumps, which would hold water within, which, in turn, would trigger spoilage reactions during storage. The method of preparation of the Palada Payasam dry mix is given in Fig. 9.8.

Fig. 9.8 Flow chart for the preparation of Palada Payasam dry mix

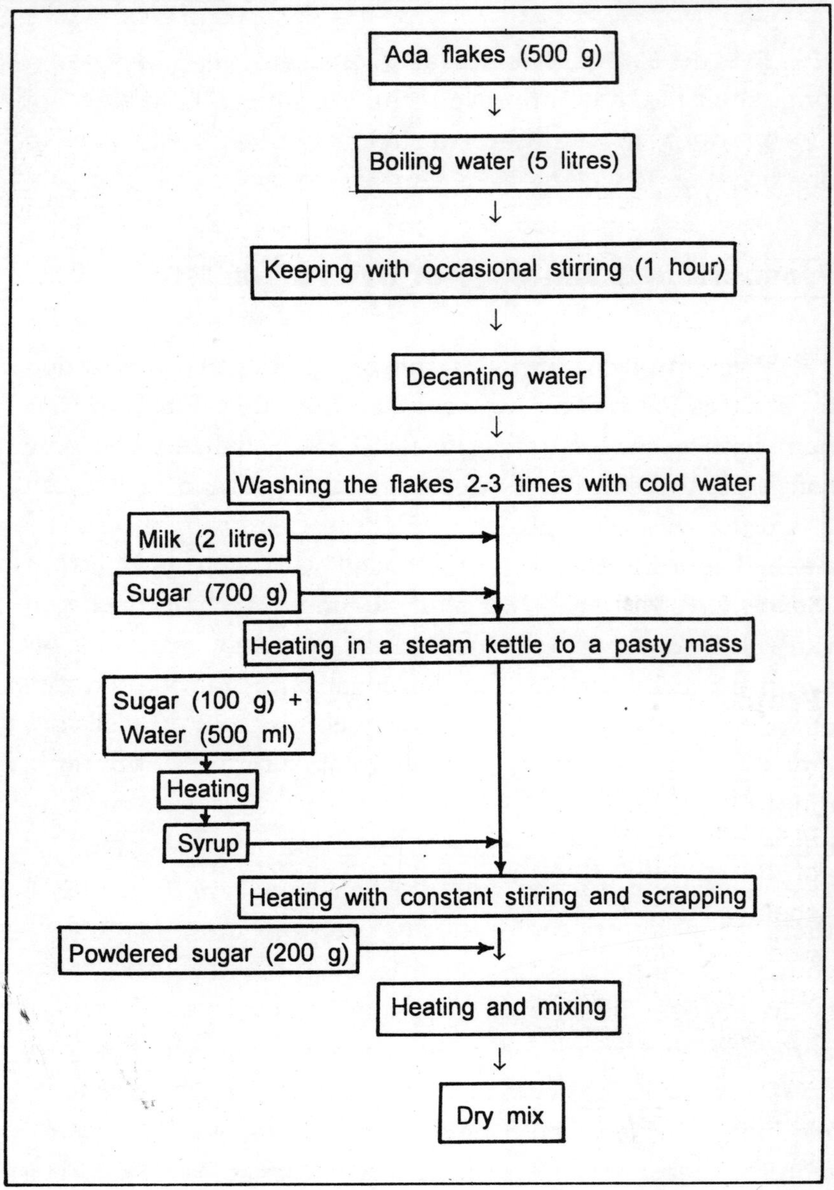

Composition

The composition of mix is given in table 9.15.

Table 9.15

Proximate Composition of Palada Payasam Mix

Constituents	% Weight
Moisture	2.5
Fat	3.0
Proteins	3.0
Lactose	3.0
Sucrose	68.0
Starch	13.0

Source : Unnikrishnan, 2003.

Preparation of Payasam with the Dry Mix

The Payasam of required consistency could be prepared by cooking 200 gms. of the dry mix in half litre of toned milk for about 12 minutes. BY this time the cooked ada would swell to attain a soft texture and would rise from the bottom of the cooking vessel and distribute uniformly throughout the body of Payasam. The reconstituted product scored better for flavour and colour than the traditional product; the ada flakes has textural quality comparable to traditional product. The composition of the product is also similar to that of the traditional product. The dry mix is well received and appreciated by the consumers.

Chemical Composition of Palada Payasam

Chemical composition of Palada Payasam is given in Table 9.16.

Table 9.16

Proximate Chemical Composition of Palada Payasam (traditional and reconstituting dry mix)

Constituents	Traditional Payasam	Reconstituted Payasam
	% Weight	
Total solids	50.8	48.5
Suspended solids	6.5	5.8
Fat	3.6	4.3
Protein	4.2	4.6
Lactose	5.0	5.2
Sucrose	31.2	27.6
Starch	4.6	4.2

Source : Unnikrishnan et.al., 2003.

Organoleptic Quality

Organoleptic scores are given in Table 9.17

Table 9.17

Organoleptic Scores of Payasam Prepared by Traditional Method and Reconstituting Dry Mix

Sensory attributes	Traditional Payasam	Reconstituted Payasam prepared from	
		Dry mix (Fresh) Score *	Dry mix (Stored for 1 year)
Colour and Appearance	8.0	8.2	8.2
Flavour	7.8	8.3	7.5
Body and Texture	7.9	7.3	7.0
Overall Acceptability	7.9	7.7	7.2

*Scores on 9-point hedonic scale

Source : Unnikrishnan *et al.*, 2003.

Shelf-life of the Dry Mix

The dry mix, developed in laboratory, may remain intact for one year when packed in pouches made of polythene pouches of 75 mm thickness and stored at room temperature. The payasam, prepared using the dry mix which had been stored for one year, was observed to have good sensory properties.

MANGO MILK POWDER

This ready-to-reconstitute dried milk product was developed at NDRI Karnal (Bhanumurthi J.L., 1977). The product contains mango pulp, milk fat, SNF, sugar and stabilizer. This product gives a refreshing milk shake on reconstitution and cooling. It can also be used in products like ice-cream, custards, jellies and shrikhand. The product has the following composition : milk fat : 12 to 14%, MSNF : 40%, Sugar 25 to 27% mango solids 14 to 18% and stabilizer 1%.

BANANA MILK POWDER

This product, also developed at NDRI Karnal, (Bhanumurthi, 1977), is made in a similar way as mango milk powder except banana is used instead of mango.

DISCUSSION QUESTIONS

1. Define Anhydrous milk fat. How is it manufactured ?
2. Define butter powder. What are the major ingredients of butter powder ?
3. Explain with the help of a flow diagram the manufacturing technology of cheese powder.
4. What do you understand by low lactose powder ? Explain.
5. How Khoa powder is prepared from buffalo milk ?
6. What are the advantages of Gulabjamun mix ?
7. Explain the production technology of Chhana powder ?
8. What is the utility of Srikhand powder ?
9. How is Kheer mix powder manufactured ?
10. Explain Palada Payasam dry mix powder.

10

Technology of Packaging

With the emergence of most uphill and challenging task of distribution of milk products, packaging has become the pivot of marketing system of the dairy industry. Today, every consumer is looking forward to a packaging material which will preserve the commodity till it is consumed. Other dimensions of packaging are its attractiveness, easy to open formula and quick disposing ability. It has a great impact on import or export of the product also, therefore, packaging recognised as the most integral part of the dairy and food industry.

Definition of Packaging

There are many definitions of packaging of which commonly quoted ones are :

- Packaging is the art, science and technology of preparing goods for transport and sale.
- Packaging may be defined as the means of ensuring the safe delivery of a product to the ultimate consumer in a sound condition at the minimum overall cost.

Expectation of Consumer from Packaging

- Easy to carry
- Non-Spillable
- Proof against ordinary delivery
- Shopping hazard
- Right amount of product
- Well presented with proper identification
- Weight marked
- Instruction for storage and use.
- Lists of ingredients.
- With all these, it must be attractive, too.

Properties to Milk Powder

The properties of dried milk product relevant to packaging are Hygroscopicity, Cohesivity, Oxygen sensitivity, Sensitivity to light, Sensitivity to heat, Bulk density, Odour pick-up, Static electricity, Bacteriological aspects.

Hygroscopicity

The principal requirement of packaging is to prevent dried products picking up moisture (Heiss & Eichner 1971). In case of whole or skim milk powder, there is a legal limit of 5% moisture at the point of sale. However,

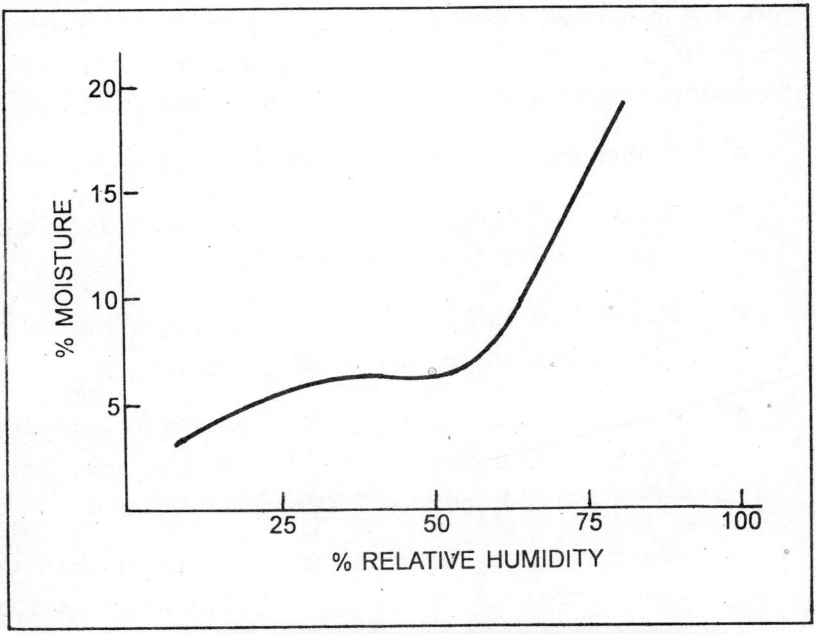

Fig. 10.1 Typical sorption isotherm for spray dried whole milk powder at 20-35°C

the manufacturer also has to allow for the delay in time between manufacture and sale and the atmospheric conditions prevailing during that time. Equilibrium Relative Humidity (ERH) takes into account the ambient conditions which may prevail. The expression of ERH is seen in the sorption isotherm (Fig. 10.1). Skim and whole milk powder exhibits the same sigmoid curve. Consequently, in conditions of high humidity, as in the tropics, packaging materials require careful selection to give satisfactory protection in terms of moisture pick-up.

Cohesivity

The cohesivity or stickiness of milk powder can be a serious problem particularly in high speed packing lines. Probably the worst powder in this respect is spray dried cheese and caseinates. Buma (1971) in the Netherlands has examined spray dried milk and has concluded that :

- Cohesion increases with decreasing particle size.
- Cohesion is independent of fat content in the range 20-45 per cent fat.
- A small amount of surface fat is sufficient to give cohesion to whole milk powder.
- Cohesion at first decreases when powder moisture increases from 2% to 4%, but from 4% onwards cohesion increases strongly.
- Cohesion in spray dried whey is higher than in spray dried whole milk of the same average particle size at a moisture content of 2-3%.

Oxygen Sensitivity

The effect of oxygen on dried milk containing butter fat has been well reported. Protection is afforded by preheating treatments which release sulphydryl compounds which function as antioxidants. Gas packing using an inert gas, i.e., nitrogen, may be practised but the pack must prevent diffusion of the gas.

Bulk Density

Bulk density on spray drying plants is controlled by feed concentration and temperature coupled to exist air temperature and the particular characteristics of the drier, especially the automizer.

In roller drying plants feed concentration and the degree of milling afterwards are the relevant factors.

Bulk density is important as packs become progressively smaller; in retail packs it can be critical. Instantized products are singularly difficult to control in terms of bulk density and the re-wet instantizing plants frequently incorporate a sizing sieve of special design and motion to give a more uniform granule (Lovell, 1972).

As a general rule, as bulk density increases free flowing properties improve.

Static Electricity

Static electricity in powder can be a problem which is difficult to resolve. Mainly, the problem is more closely associated with packaging materials, e.g., polythene. Manufacturers of polythene film incorporate polyethylene glycol alkyl esters which function as anti-static agents in their films (Brydson, 1969).

Sensitivity to Light

Normally, light is not a problem with dried milk, sunlight will bleach the surface of milk powder and accelerate fat oxidation in fat containing powder.

Sensitivity to Heat

Milk powder is sensitive to heat which can affect their flavour and solubility. Ideally, milk powder should be kept in cool and dry atmosphere, certainly less than 30°C.

Odour Pick-up

Product containing fat are prone to odour pick-up and butterfat is well-known in odour pick-up quality. High fat powder is particularly susceptible to pick up not only from packaging materials, but also from the surrounding atmosphere in stores and warehouses. Several instances of taint due to odour from Kraft paper, polythene film, fibreboard cases and, in one instance, from the rubber sealing compound in a steel drum have been reported. (BSI PD 6459 : 1971)

Bacteriological Aspects

Products of normal ERH do not give rise to bacteriological problems.

Packaging Material

The requirements of packaging materials for dried milk products are :

- The material should have a high resistance to the passage of moisture.
- It should have a high resistance to the passage of gases.
- It should have adequate mechanical strength to withstand handling, transport and storage.
- It should be resistant to extremes of temperature and humidity.
- It should be light and easy to handle.
- It should not impart any odour or flavour to the product.
- It should be satisfactory on any packaging machine designed to handle the material.

No single material can fulfil all these requirements and it is necessary to make a balance.

PACKAGING MATERIALS COMMONLY USED FOR DRIED MILK & DRIED MILK PRODUCT

Tinplate

Tin plate has been used for many years and is still employed for export and retail packs. Today, there are several choices available : standandard tin plate, light weight tin plate, double reduced tin plate, etc. It fulfils very many of the requirements and perhaps only the disadvantages require mention. These are High cost, storage space required for empty container. Weight, although this is becoming less with double reduced tinplate, corrosion hazard in high humidity conditions.

Paper and Paperboard

A very considerable portion of packaged foods is stored and distributed in packages made out of paper or paper based materials. Paper and paperboard contribute to mechanical/strength (stiffness and sturdiness).

Advantages—Low cost, easy availability and versatility, paper is likely to retain its predominant position in packaging industries.

Disadvantages are—Paper is highly permeable to gases, vapour and moisture and lose its strength when wet. Ordinary paper is not grease and oil resistant but can be made resistant by mechanical processes during manufacturing.

Paper Manufacturing

Paper pulp is produced from wood chips by acid or alkaline hydrolysis. The pulp is suspended in water and beaten with rotating impellers and knives to split the cellulose fibres longitudinally. The fibres are then refined and passed through heated rollers to reduce the moisture content and then through finishing rollers to give the final surface properties to the paper. Alkaline and acid hydrolysis produces sulphate pulp. Sulphate pulp is used to make kraft paper. It is of Scandinavian origin and means 'strong'. This is strong paper and it is used for manufacture of corrugated fibre board box.

Grease proof paper is made from sulphite pulp in which the fibres are thoroughly beaten to produce a closer structure. It is resistant to oils and fats. But, this property is lost when the paper becomes wet. It is widely used for dairy products.

Thicker paper is called paper board. The lightest standard board is 0.19 mm. thick and heavy paper is of 0.125 mm. thickness. Paper boards are used for carton making. The WVTR and GTR of paper board are very high and it needs to be laminated with plastic polymers or coated with material(s) which would reduce the same.

Cellophanes

Cellophane was the first commercial film uncoated cellophane is flexible, strong, transparent, grease proof, hygroscopic and highly permeable to water vapour and low permeable to odours.

Cellophane Manufacturing

It is manufactured from highly purified cellulose derived from bleached sulphite pulp. By incorporating various coating and modifications over 100 different grades of cellophane are available now.

Use

Cellophane is widely laminated and singly used as overwraps for cartons. Adhesive sealing is generally used.

Polyethylene

Polyethylene has found wide application in packaging.

Advantages

It has good flexibility even at low temperature :

- Transparency
- is a good moisture barrier.
- has excellent heat sealing properties.

Its resistance to the passage of gases is related to its density. Low density being more permeable than high density polyethylene.

Disadvantages

Its proneness to static electricity which can be a problem on bag making and/or filling equipment. This is overcome by the incorporation of anti-static substances.

Use

It is an important component when used in conjunction with other materials either as a coating or as a laminate.

Polyethylene Manufacturing

It is made by polymerisation of ethylene which, in its natural state, is gas. Polymerisation under very high pressure (1200 atm.) and higher temperature (150-200°C) in presence of traces of oxygen yields a macromolecule with very high degree of branching, whereas at low-pressure forms polymers with long parallel liner chains (In case of high density polyethylene). Because of the branched chain structure, close packing of the polymer chains is not possible. Therefore, the density of the finished product is low and the product is called low density polythene (LDPE).

Comparison of barrier and resistant properties of plastic packaging material is given in Table 10.1.

Polypropylene

Polypropylene is very similar in many respects of high density polyethylene and differs, in that it has a higher softening temperatue and in the oriented form has a higher tensile strength and improved gas barrier properties. It also has less opacity than polyethylene and can be made as clear as cellophane film.

Polyester Films

Polyester can be produced by reaching ethylene glycol with terephthalic acid. Polyester films have a high tensile strength, are resistant to heat and have good gas and moisture barrier properties. They are used in laminates

Table 10.1

Barrier and Resistant Properties of Plastic Packaging Materials

Property	LDPE[1]	HDPE	PET[23]	PP	PS	PA	PVC	EVP[4,5]	PVDC
Barrier to Moisture	■	■	■	■	●	■	■	✧	■
O_2	O	O	●	O	O	■	●	■	■
CO_2	O	O	■	O	O	■	●	■	■
Aroma	O	O	●	O	O	■	●	■	■
Resistance to Oils and Fats	O	■	■	●	●	■	■	✧	✧
Pasteurisation	✧	■	■	■	●	■	●	●	●
Sterilisation	✧	O	■	■	✧	●	✧	●	●

■ Very good ● Good O Slight ✧ Poor

1. Lamination of LDPE or HDPE with EVOH, EVA, OPP, PVDC and/or PA besides the PET enhances the barrier property for aroma constituents. Multiple barrier films considerably decrease the loss of aroma constituents.
2. Oxygen permeability through PET decreases with metallisation. With an increase in the metallisation from 0.0 to 39.4 microns, O_2 permeability decreases from 77.5 to 0.26 per cm^3 – mil/(24 h-m^2-atm) (Dupont co. 1990). Similarly, when metallisation thickness is 15.24 microns oxygen permeability of PA decreases from 4.85 to 1.09 cm^3– mil/(24 h-m^2-atm).
3. EVOH (ii) PVA (iii) PAN (polyacrylonitrile) and (iv) PVDC provide a better barrier to oxygen than PET and the extent is in the order started.
4. Water-soluble.
5. By copolymerising EVOH with ethylene, water solubility decreases but barrier properties are retained.

Source : Techline Ind. Data Service Ltd. (1998).

and their disadvantages would appear to be their high cost and the difficulty in heat sealing.

Polystryene

Manufactured from ethylene and benzene, which are cheap. Polystyrene film has poor moisture and gas resistance even in oriented films. It is difficult to heat seal and tends to run or stick on the heat sealing jaws. It does have a very high degree of clarity and, hence, its use is mainly confined to overwrapping.

Polyamides

These are more commonly known as the nylons of which Nylon 6 and 11 find the widest application. It is a strong material a good gas barrier and welds easily. The moisture barrier properties of Nylon 6 are poor but those of Nylon 11 are good. The latter is capable of a high degree of clarity. Most nylon packaging films produced in USA are from Nylon 6 while, European films are usually Nylon 11, due to lower raw material costs films from Nylon 6 have higher temperature, grease and oil resistance than Nylon 11 films.

Co-polymer of Vinyl Chloride and Vinylidene Chloride

More commonly known by commercial names of Saran and Viclan, these have a high degree of clarity and good gas and moisture barrier properties.

Properties of Flexible packaging materials are given in Table 10.2.

Table 10.2

Properties of Flexible Packaging Materials

Packaging Material	Standard abbreviations	WVTR at 88.8°F (38°C) 90% RH g/m²/24h	GTR at 77°F (25°C) cc/m²/ 24 h atm	Softening point (Vicat)°C
1	2	3	4	5
Polyethylene	PE	16-23	7000-8000	85-87
Low density Polyethylene	LDPE			
Linear Low density				
Polyethylene	LLDPE			
Medium density	MDPE			
Polyethylene				
High density Polyethylene	HDPE	5-11	15,00-2000	120-130
Poly propylene	PP			
Cast polypropylene	CPP	7-9	2000-3000	150
Biaxially oriented				
Polypropylene	BOPP	4	2000-2500	
Polyethylene trephtalate	PET	21-30	50-90	250
(Polyester)				
Polystyrene	PS	100-155	4000-5000	78-103
Oriented Polystyrene	OPS			
High impact polystyrene	HIPS			
Expanded or foamed polystyrene	EPS			
Polyamide (Nylon)	PA			
Nylon 6-polycimide		250-350	40	216 (MP)
Nylon 11-Polyamide		60-100	400	216
Poly vinyl chloride	PVC	30-40	300-400	82
Poly vinylidene chloride (SARAN)	PVDC	2-4	5-25	80
Poly vinyl acetate	Pva			

(Contd.)

1	2	3	4	5
Polyvinyl alcohol	PVAL			
Polycarbonate	PC	50-150	3000	165
Ethylene vinyl acetate	EVA	20-50	10000	95 (M.P.)
Ethylene vinyl alcohol	EVOH	20-50	0.5-20	
Cellulose acetate	CA	100-200	2000-3000	
PT cellophane	10-12	10-12		
MST Cellophane**	4-8	3-9		
MXXT Cellophane**				
Cellulose acetate-butyrate	CAB			
Acrylonitrile butadine	ABSI	80-150	750-1000	90

Thickness : 100 gauge = 1 mil. = 25.4 micron = 0.0254 mm. (Values found using 25 micron film).

** Coding system for regenerated cellulose films.

A = anchored (the coating of the film is anchored which makes it moisture resistant).
C = Coloured; D = demi (half) (moisture proof coating on one side only);
L = Less mositure-proof than standard; M = Moisture-proof;
P = Plain (without coating, not moisture – proof and not heat sealable;
S = Sealable by heating; T = transparent; not coloured X = polymer coated (PVDC, Saran)

Source : Balasubrahmanyan (1995).

Aluminium Foil

Aluminium foil is defined as a solid sheet section rolled to a thickness less than 0.006 inches. It is impermeable to water vapour and gases. Aluminium has excellent properties like reflectivity, emissivity, thermal conductivity, light weight, corrosion resistance, workability, grease and oil resistance, tastelessness and odourlessness, heat and flame resistance, opacity and non-toxicity. It will not support bacterial or mould growth. Its use as far as dried products are concerned is in laminates, where it is used in thickness of 0.00030 upto 0.0007 inches. There is a limited application as a diaphragm in lever lid tins.

Laminates

The range of laminates is considerable and it is not easy to select the right laminate from the bewildering array. In general terms, laminates have the additive properties of the materials involved.

Glass

In the case of the glass jar as a container for dried products, it is only as good as its closure. As for the main body of the jar, the glass must be coloured to prevent bleaching and flavour defects. The principal problem is of weight and the resistance to damage compared to other types of containers.

STORAGE OF PACKAGING MATERIAL

Care needs to be exercised in the storage of packaging materials. Store should be clean and dry. Films on reels should be supported by a rod through the core resting on a trunnionfufestation must also be guarded against, while stock rotation should eliminate many of the problems.

Packages

Bulk Storage

(i) *Silos* : Silos are short term containers for powder, the principal objection to long term storage being compaction and lumping. Powder is stored in silos to facilitate packing in one shift. Silos are of stainless steel construction and available of upto five tons capacity. Powder being cooled to 70°F before filling in silo. Cooling is essential if compaction is to be minimized and particularly if whole milk powder is stored in bulk. It has reported on bulk storage, recommends limits of 3 per cent moisture for whole milk powder and 5% moisture for skim milk powder at RH of 35%.

(ii) *Multiwall paper sacks* : Multiwall paper sacks (IS : 1165-1975) are most familiar form of bulk milk powder container. In general, the weight of paper, number of thickness, nature of liner, weight of coating should be related to the conditions which sack will experience.

The sack itself falls into two main types : Open mouth and valve sacks. Both, in term of material, have 3 to 6 layers of paper. The multi wall bags can be tested for their transport-worthiness according to IS : 7028.

(iii) *Flexible bulk packs* : Pillow packs either from "lay flat" tubing or from the sheet are used.

Other packs are Fibre board, metal drums, tins etc.

Retail Packs

(i) *Tin Plate cans* : The lever lid can is common place and incorporates either a paper or foil tamper proof diaphragm. It is almost universally used for instant skim milk granules and baby foods. The important features of this container are the ring and lid components and it is here that there is a danger of moisture ingress. If these do not make adequate contact. This is particularly the case when a paper diaphragm is incorporated but not so when a foil diaphragm is used. In either case, it is important to ensure that the end component is double-seamed. A foil diaphragm lever lid can is being used for an instant whole milk powder. This also employs gas packing.

Can lids are sealed by a double seam in a seaming machine (or seamer). The 'first operation roller' rolls the cover hook around the body hook (Fig. 10.2 (a)) and the second operation roller then tightens the two hooks to produce the double seam. (Fig. 10.2(b)). A thermoplastic sealing compound melts during heat processing and fills the spaces in the seam to provide an additional barrier to contaminants. The can seam is the weakest point of the can and the seam dimensions are routinely examined by staff to ensure that they comply with the specifications.

(ii) *Collar Cans* : It is used only for export. This employs a wind-off metal strip running round the circumference of a body component.

(iii) *Lined Cartons* : Most of the carton packs are formed from one aluminium foil paper-polythene laminated inside a carton. Long term storage trials at 90°F and 95% RH for over 9 months have shown that the increase in moisture content is under 1 per cent on an average.

(iv) *Plastic bottle* : Another alternative is a plastic bottle with screw top lid and inner sealing membrane plastic bottles of HDPE are being tried for 200 g and 400 g of powder.

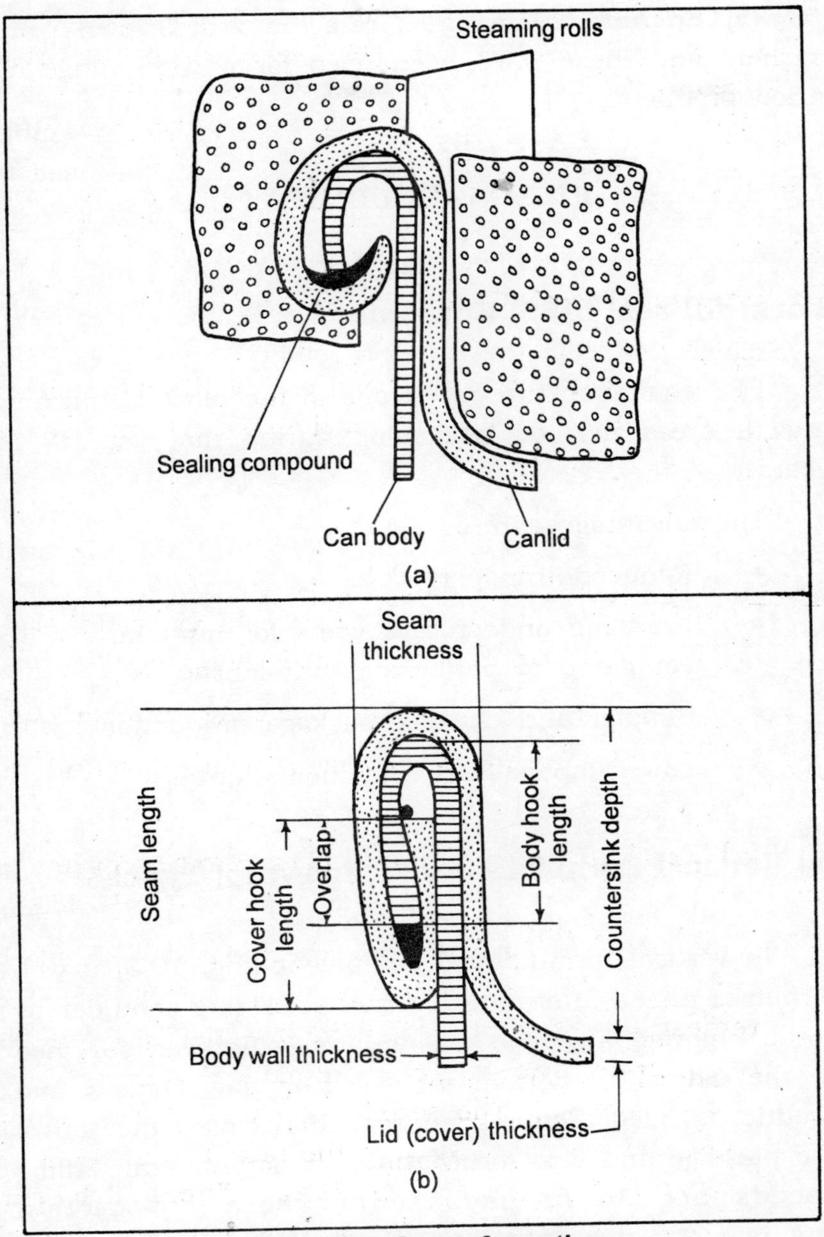

Fig. 10.2 Can Steam formation

(a) First operation, (b) Double Seam (After Hersom and Hulland 1980)

(v) *Sachets* : Laminates came into their own with sachets and these have been used for a wide range of dried products.

FILLING EQUIPMENT

Form-fill-seal (FFS) Equipment

FFS equipment has been one of the most significant growth areas in food packaging during the last 10-15 years.

The advantages are :

- Reduced transport
- Handling and storage costs for material compared to performed containers
- Simpler and cheaper package production
- Lower labour cost and higher output.

(a) Vertical Form-fill-seal (Transwrap) Types

In vertical form-fill-seal equipment (Fig. 10.3) a web of film is pulled intermittently over a forming shoulder by vertical movement of the sealing jaws. A fin seal is formed at the side. The bottom is sealed by jaw sealers and product is filled. The second seal then closes the top of the package and also forms the next bottom seal. Filling speeds are 30-90 packs/min. There are certain requirement which must be met with this type of machine.

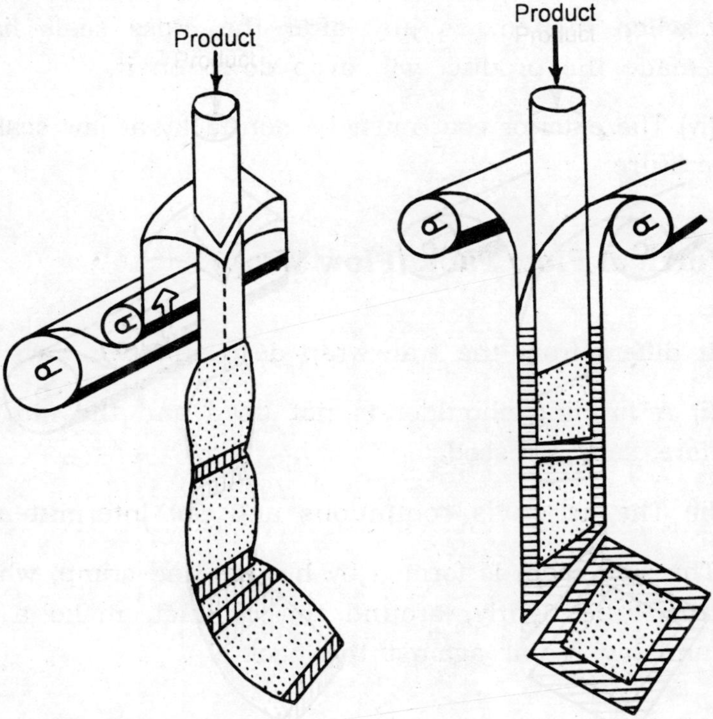

Fig. 10.3 Vertical form-fill-seal (trans wrap)

(***Courtesy*** : Flex Industries Ltd.)

(i) The packaging material must be thin and flexible, cellulose film laminated to itself or polythene for example are satisfactory. Aluminium foil laminates are not very satisfactory.

(ii) The packaging material must have good slip characteristcs and resistance to creasing or cracking in order to pass over the filling tube.

(iii) A high melt strength to support the product on the hot seal, and the heat seal coating must be very tacky when molten, as just after the cross seals have been made the product will drop down on it.

(iv) The exterior coat must be non-tacky at jaw sealing temperature.

(b) Vertical Flow Pack (Flow Wrap)

It differs from the transwrap design in two ways :

(i) A forming shoulder is not used and the film is, therefore, less stressed.

(ii) The action is continuous and not intermittent.

The side seam is formed by heaters and crimp, which pull the tube tightly, around the product, make a fin seal and lay it flat against the pack.

(c) Horizontal Form-fill-seal (HFFS) (Pillow pack of flow wrap)

This equipment has gained in popularity due to greater speed (upto 600 packs/min.) and flexibility.

A modification of this equipment is used to fill laminated cartons aseptically. A web of material is sterilised in a bath of hydrogen peroxide and formed into a vertical tube. An internal heater vapories any remaining hydrogen peroxide. The tube is then filled, sealed through the product shaped into a carton and top sealed. The 'ears' on the base of the carton are folded flat and sealed into place.

Sachet Pack Machines

These are of two main types : Vertical machines and Horizontal machines.

In sachet pack machines, either horizontal or vertical packs are formed from single or double sheets of film. Horizontal single-web machines fold the film over a triangular shoulder and then form two side seams (Fig. 10.4) The sachets are then separated, open by a jet of compressed air, filled and heat sealed across the top. The vertical single web machine is similar to transwrap machine. Horizontal machines have a smaller distance for the product to fall into the package and vertical machines have lower cost and take up less floor space.

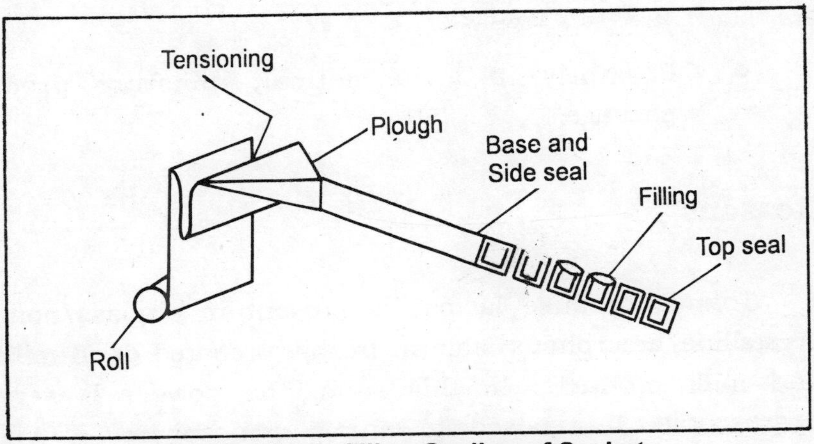

Fig. 10.4 Form-filling-Sealing of Sachets

(***Courtsey*** : Engelmann and Buckham Machinery Ltd.)

On two-reel machines, one web forms the front and the second forms the back of the pack. Two blanks are cut from a roll of film, brought together and sealed on three sides. The package is filled and the final seal is made. Sachet machines are widely used for powders/ granules, liquids/sauces. Filling speed is 70-1000 min^{1} depending on the size of the sachet and type of product. Sachets are automatically cartoned and cartoning machine is, therefore, an integral part of sachet production (Guise, 1987a, and 1987b).

Requirement Which Must be Met during Powder Packaging

Powder packaging should be carried out :

- in a dry atmosphere between the drier and sealed package.
- Promptly in a hermetical, moisture proof package.

Reasons

This is because lactose is present in a glass/non-crystalline/amorphous state in freshly prepared dried milk and milk products. In this state, the powder is very hygroscopic. If exposed to humid atmosphere, it will readily absorb moisture, become sticky and cake in storage.

Packaging of Whole Milk Powder and Infant Milk Powder

Since the whole milk powder and infant milk powder are rich in fat (20-30%) packaging material should provide protection against oxygen and moisture. Generally, tin plate cans are used. Gas packing is the most practical commercial procedure for increasing the shelf-life of whole milk powder. By reducing the oxygen content to 2% or less (by volume) in the head space gas, the shelf-life is roughly doubled.

The general procedure is to immediately remove oxygen by subjecting the product to a vacuum of 71 cm. (28 in) of Hg within 24 hours of drying, with final packaging within a few days.

To obtain a low level of headspace-oxygen in whole milk powder, a double gassing technique is applied. The customary procedure is to collect filled, air-tight metal cans (with small brogue hole) on trays to be conveyed into the vacuum chamber/cabinet. The air is removed rapidly (in 60 secs.) under a very high vacuum of 74 cm. (29 in) of Hg. After a 2-5 minute hold, the pressure is restored with nitrogen of 0.5 to 1.0 psi above atmospheric pressure. Nitrogen may be replaced with a mixture of nitrogen and carbon-dioxide, the latter being restricted to 5-20%. Once removed from the chambers, the containers are sealed by soldering the 1-2 mm. holes on the lid. The containers are then held for oxygen desorption. When an oxygen equilibrium has been attained in the head space, usually within a week, but at the most within ten days, the cans are punctured and the vacuum treatment pressure restoration with nitrogen and sealing steps are repeated.

Packaging of Skim Milk Powder

Skim milk powder has very low fat content and as such, a good moisture barrier material is adequate for short-term storage. For industrial use and storage SMP may be packed in barrels, drums and bags. Multi-wall paper sacks with one polythylene liner (IS : 1165-1975) is used for bulk packaging. The multiwall bags can be tested for these transport–worthiness according to IS : 7028. Some skim milk powder has also been sold in polythylene bag of 1 kg. Even plastic bottles of HDPE are being tried for 200 g and 400 g of powder.

Packaging of Malted Milk Food

The Spray dried malted milk food packaged in metalized polyester laminate pouches has shelf-life of more than 1 year at 25°C.

LABELLING

The label is the primary point of contact between a processor and a consumer and is, therefore, an integral part of the marketing strategy for a product. The label is the main method of persuading a purchaser to buy a product without having sampled it, rather than a competing brand on a retail shelf. Details of the factors to be taken into account in the design of labels, legislative requirements and the information required. Labels are made from paper, plastic film, foil or laminated materials, pre-printed by either lithography or rotagravure techniques. Cans and glass bottles are usually labelled using a hot-melt adhesive at 500 or more containers/minute. A wide variety of label combinations are possible to produce a distinctive brand image for the product. (Blanchfield, 2000).

The following particulars are expectd to be marked or labelled on each container of dried milk and milk products :

- Name and type of the materials (Trade/Brand name if any)
- Name of manufacturer and address
- Batch or code number
- Net weight
- Process of manufacture (Spray or roller dried)
- Date of manufacture and/or expiry date.
- Nutrition information (Baby food & health drink).
- Direction for use
- Feeding Table (Baby food)
- Warning/Caution Reminder (if any) etc.

BAR CODES AND OTHER MARKING

The Universal Print Code (UPC) or bar code is printed on consumer packs for laser reading at retail checkouts. It avoids the need for individual price labelling of packs and allows itemised bills to be produced for customers. The bar code information is fed to a computer which deducts the item from the store inventory and, thus, enable faster stock-taking, detection of pilferage and automatic re-ordering. The information is also collated into product sales reports, which can be used by store managers to adjust the shelf space allotment to specific items, or produce data on competitor's sales or the result of promotion and marketing strategies. Corrugated board shipping containers are also bar coded to inform the carrier about the destination but this is not yet possible directly into shrink-wrapping (Osborne 1986).

Markings are also required on containers to show the 'shelt by' or 'use by' date in many countries. A manufacturer's code is printed onto containers to identify the factory, the production line and the shift during which the product was made coding lasers are non-contact and produce permanent marks without the use of inks and solvents. They are fully programmable to easily change the characters, and are capable of producing 400-2000 characters/second on paper board, metal, glass, plastic and foil (Campey, 1987).

Packaging of evaporated and Sweetened condensed Milk

Packages

Cans : The evaporated milk container must withstand the pressure changes in the process of sterilization. There are two types of cans.

(i) The Sanitary type (the can with solder seal) The sanitary cans are open at the top and provided with separate lid.

(ii) The vent hole cans : These are used for evaporated milk. These cans have a small orifice for filling the product and are sealed by tipping the vent hold with a drop of solder.

Aseptic paperboard Packages : Now-a-days, it is also possible to pack sweetened condensed milk in aseptic paperboard packages.

Beech-wood barrels : Sweetened condensed milk is also packed in beach-wood barrels, holding about 300 kg, for supply to large-scale users.

Leak detection for cans : The cans are passed through a hot water bath. In case of leaking of can, the pressure inside expels some of the excess air causing visibility of air bubbles. Such cans are picked off and returned to the can inspection for repair.

I. Packaging of Evaporated Milk

The vent hole cans are used for evaporated milk Filling is carried out at low temperature to avoid foam formation. The cans should be filled as nearby full as possible without fouling the surfaces of the tin or around the vent hold causing interference with the solder. If cans are filled too full it may cause flipping because of limited air space to allow for unavoidable expansion.

When filling vent hole cans there is a tendency for the milk to wet the surface around the orifice. This may interfere with proper functioning of the solder. Hence, the top of the can should be clean and dry. The filling machines have automatic multiple piston pumps.

The filled and sealed cans pass from the filling machine to autoclave.

UHT treated condensed milk, after cooling, is packed in aseptic paper board packages and stored.

II. Packaging of Sweetened Condensed Milk

Sterility of Cans

Cans used for filling sweetened condensed milk should be sterilized before use, otherwise they may contribute to the microflora. Lids should also be sterilized properly before use and cans should be regularly checked for microbiological quality by rinse method. The cans must be filled as full as possible as minimized head space otherwise aerobic organisms, if any present, may grow and cause defects in the product. Imperfect seals or can damage may permit the entry of the organisms as well as of air which favours the growth of micro-organisms.

Packaging

The sanitary cans are open at the top and provided with separated lid. After filling SCM as full as possible, the lid is placed and the edges seamed by machine. A gasket cement in the peripheral rim of the lid assists in providing dependable hermetic seal. The contamination between storage tank and sealing machine must be avoided.

SANITARY CONDITION OF THE SEALING ROOM

The sealing room must be kept in sanitory condition, clean and free from accumulation of rubbish. It should be closed to the outside and visitors. Filtered air should be used in the filling and sealing room.

STERILITY OF EQUIPMENT

- The equipments used should be sterilized by use of chlorine, steam or other sterilants.
- The pipelines, filling machine should be thoroughly washed, cleaned and sterilized.
- Sterility can be checked by rinse or swab method.

- Gaskets and values are a major hazard. Gaskets should be dislodged from the sockets, cleaned and immersed in a sterilant.

Valves often form blind pockets. The smear of SCM lies in the seat of the valve and get dislodged when moved. This comes in contact with the product. Care should be taken to see the valves fitting etc. Satisfy the requirements of maintenance under hygenic condition.

The filling machine should be emptied after a day's run; all detachable parts to be dismantled washed, scabbed, steam dried and replace in the machine.

The filling machines should be hooded when not in use to protect them from dust, chance insects and other agencies of contamination.

DISCUSSION QUESTIONS

1. Define packaging. What are the expectations of consumer from packaging ?
2. What are the packaging relevances of milk powder ?
3. Explain the packaging material commonly used for dried milk and products.
4. Describe filling equipments in detail.
5. How is malted milk powder packed ?
6. Explain packaging method of sweetened condensed milk.

Question Bank

1. Condensed milk first invented by
   (a) Luis Pasteur (b) Gail Borden
   (c) **Nicholas Appert** (d) Charles Page
2. First commercial production of condensed milk in India was in the year :
   (a) **1961** (b) 1965
   (c) 1955 (d) 1962
3. First dairy that manufactured SCM in India was :
   (a) Model Dairy (b) Modern Dairy
   (c) Vishakha Dairy (d) **Amul Dairy**
4. The minimum fat level in USCM is :
   (a) 7% (b) **8%**
   (c) 9% (d) 10%
5. The minimum Sugar % in SCM is :
   (a) 35% (b) 41%
   (c) 42% (d) **40%**

6. Specific gravity of SCM at 16°*C* :
   (a) 1.282 (b) **1.303**
   (c) 1.083 (d) 1.355
7. The cooked flavour in SCM is due to :
   (a) **Sulfydryl compounds** (b) bacteria
   (c) Molds (d) Overheating
8. Maillard-type browning in SCM is due to :
   (a) Interaction of fat and carbohydrate
   (b) Protein and Fat
   (c) **Casein with Lactose**
   (d) Minerals with Sugar
9. Number of Standardization required in SCM manufacture is :
   (a) 2 (b) 4
   (c) **3** (d) 1
10. Forewarming Temperature for SCM is :
    (a) 71-81°C (b) **82-93°C**
    (c) 62-73°C (d) 80-90°C
11. Entrainment seperator is used for :
    (a) Milk seperation
    (b) **Reclaimation of milk particles**
    (c) To condense the milk
    (d) To heat the milk.
12. Vacuum pump is used for :
    (a) Condensing of milk
    (b) Production of vacuum
    (c) **Production and maintaining partial vacuum**
    (d) To reduce pressure

13. Striking the batch means :
    (a) **Reaching to a correct concentration**
    (b) Heating the batch
    (c) Shut down the heat source
    (d) Cooling the batch
14. Seeding in SCM refers to :
    (a) Cooling
    (b) heating
    (c) **Introduction of lactose in powder form**
    (d) addition of casein.
15. SCM should be stored at :
    (a) 5°C (b) 8°C
    (c) **10°C** (d) 12°C
16. Mold buttons in SCM in due to
    (a) Torula lactis (b) Str. Lactis
    (c) **Asperigillus Ripens** (d) Penicilium.
17. Bloats in SCM is connected with :
    (a) Asperigillus (b) **Torula lactis**
    (c) Shygella (d) Yeast
18. Sandiness in SCM is due to :
    (a) Sand particles (b) **Excess sugar**
    (c) Low Sugar (d) Low protein.
19. Age thickening is related with :
    (a) **Excess temperature** (b) yeast
    (c) Bacteria (d) Molds
20. Feathering in coffee is due to :
    (a) Poor Coffee Powder (b) Excess milk
    (c) **Excess viscosity in SCM** (d) High heat

21. The protein content in buttermilk powder is
    (a) **34.3%** (b) 32.5%
    (c) 33.2% (d) 34.8%
22. Whey powder is basically used for
    (a) Lassi (b) Poultry and livestock feed
    (c) Sweets (d) **Breads**
23. The fat percentage in cream powder is usually :
    (a) 30% (b) 40%
    (c) **65%** (d) 80%
24. Butter powder is used for making :
    (a) Butter (b) Cream
    (c) Ghee (d) **Ice-cream**
25. The sugar % in ice-cream milk powder is :
    (a) 25% (b) 30% (c) 35% (d) **37%**
26. Cheese powder should be stored at :
    (a) 5°C (b) 7°C (c) **10°C** (d) 12°C
27. Malted milk powder was first invented by :
    (a) Nicholas (b) Borden
    (c) **William Horlick** (d) Henry
28. Malted powder basically used for :
    (a) Beverages (b) **Infant feeding**
    (c) Bakery products (d) Chocolate
29. Minimum milk fat in malted powder is :
    (a) 5% (b) 6% (c) **7%** (d) 8%
30. Malted powder is generally prepared from :
    (a) Rice (b) **Wheat**
    (c) Jowar (d) Bajra

31. Infant milk food is an ideal food for the children :

(a) below 1 year (b) **below 2 years**

(c) below 3 years (d) below 5 years

32. Baby Food should contain minimum :

(a) 30% Carbohydrate (b) **35% Carbohydrate**

(c) 40% Carbohydrate (d) 45% Carbohydrate

33. The word fortification means :

(a) addition of major

(b) **addition of aminoacids and vitamins**

(c) addition of proteins

(d) addition of sugar.

34. Ferrous sulphate is used in baby food for :

(a) Bone formation (b) **Blood formation**

(c) Muscle building (d) Teething

35. Dry sodium caseinate is used as a :

(a) Food (b) **Dairy Whitener**

(c) Bread manufacture (d) Chocolate

36. % lactic acid in Srikhand powder is :

(a) 0.30 (b) 0.35

(c) **0.25** (d) 0.20

37. The base material for Srikhand powder is :

(a) **Chakka** (b) Curd

(c) Srikhand Wadi (d) Milk

38. The protein content in chhana powder is :

(a) 40% (b) 42.3%

(c) 45% (d) **46.3%**

39. % lactose in Khoa powder is :

(a) 20.5% (b) 25.6% (c) **30.6%** (d) 35.2%

40. Shelf-life of Khoa powder is :

(a) 15 days (b) 30 days

(c) **90 days** (d) 120 days

41. Storage temperature of evaporated milk is :

(a) 5°C (b) **16°C** (c) 18°C (d) 10°C

42. Pilot sterilization test is done :

(a) **To determine the amount of stabilizer**

(b) To determine the amount of emulsifier

(c) To determine the acidity

(d) To check the sterilization of the product

43. Shaking of evaporated milk refers to :

(a) Stirring (b) **Breaking down the curd**

(c) Mixing (d) Checking the viscosity.

44. Salt balance theory refers to :

(a) acidity of milk

(b) pH of milk

(c) Ash percentage of milk

(d) **Heat stability of milk**

45. Homogenization temperature for evaporated milk is :

(a) 40°-50°C (b) 30-40°C

(c) **50-55°C** (d) 35-40°C

46. For fortification is evaporated milk :

(a) Lactose used

(b) **Vitamin D concentrate is used**

(c) Iron used

(d) Minerals used

47. Evaporated milk is cooled at :

(a) 20°C (b) 25°C

(c) **27°C** (d) 15°C

48. Packaging temperature for evaporated milk is :

(a) 10°C (b) 8°C

(c) **5°C** (d) 4°C

49. Browning in evaporated milk is due to :

(a) Bacteria (b) Yeast

(c) Mold (d) **High heat**

50. Gassy defect is evaporated milk is due to :

(a) **Low temperature** (b) high heat

(c) Molds (d) Yeasts

51. Dried milk is preferred in the dairy industry because of :

(a) **Shelf-life** (b) High Protein content

(c) Occupy less space (d) low cost

52. In dried milk nutrients are :

(a) lower than milk (b) higher than milk

(c) **same as raw material** (d) very high

53. Dried milk first invented by :

(a) **Grimwade** (b) Nicholas Appert

(c) Percy (d) Stauf

54. Milk powder should contain not more than :

(a) 3% moisture (b) 4% moisture

(c) **5% moisture** (d) 6% moisture

55. Minimum fat level in milk powder is :

(a) 25% (b) **26%**

(c) 24% (d) 22%

56. Skim milk powder should not contain more than :
(a) **1.5% fat** (b) 2.5% fat
(c) 2% fat (d) 1% fat

57. Maximum moisture content in skim milk is :
(a) 4% (b) **5%** (c) 3% (d) 2%

58. Average lactose content in whole milk powder is :
(a) 37.3% (b) **38.2%**
(c) 36.2% (d) 35.3%

59. Average protein in Skim milk powder is :
(a) 34.8% (b) 35.2%
(c) **35.9%** (d) 36.8%

60. Average ash content in Skim milk powder is :
(a) 7% (b) **8%** (c) 9% (d) 6%

61. WMP is a good source for :
(a) Vitamin B (b) Vitamin C
(c) Vitamin D (d) **Vitamin A**

62. Caramelized flavour in milk powder is due to :
(a) Protein (b) fat
(c) **Sulfydryl compound** (d) denaturation

63. Draw dryer is made of :
(a) Copper (b) Iron
(c) **Steel** (d) Nickel

64. Heat classified SMP are generally of :
(a) Two types (b) **Three types**
(c) Four types (d) Five types

65. Vacuum dryer maintains a vacuum pressure of :
(a) 65-75.5 cen Hg. (b) 63.2-69.5 cen. Hg
(c) **68.5-73.5 cen. Hg** (d) 64.5-67.5 cen. Hg.

66. The average speed of the drum in drum dryer is :
    (a) 15-20 rpm (b) 22-24 rpm
    (c) 12-14 rpm (d) **14-19 rpm.**
67. The ratio of milk pre-condensation in dry drying is :
    (a) 3 : 1 (b) **2 : 1**
    (c) 1 : 2 (d) 2 : 4
68. Forewarming temperature in drum dryers is :
    (a) 83°C/10 min. (b) **85°C/10 min.**
    (c) 84°C/12 min. (d) 82°C/15 min.
69. Pasteurization temperature for low heat SMP
    (a) **71°C/15 sec.** (b) 75°C/10 sec.
    (c) 68°C/15 sec. (d) 63°C/30 min.
70. Cooling temperature for low heat SMP is :
    (a) 30°C (b) **32°C**
    (c) 34°C (d) 35°C
71. Pasteurization temperature for high heat SMP is :
    (a) 75°C/10 sec. (b) 72°C/15 sec.
    (c) **71°C/15 sec.** (d) 73°C/10 sec.
72. Spray drying temperature for high heat SMP is :
    (a) 75°C/15 sec. (b) 82°C/10 sec.
    (c) 80°C/12 sec. (d) **85°C**
73. Preheating temperature for WMP manufacture is :
    (a) 75°C (b) **71°C**
    (c) 63°C (d) 65°C
74. Sifting in powder manufacturing means :
    (a) Transportation (b) Straining
    (c) **Separation of different particles**
    (d) Lifting of powder particles.

75. WMP sifting is done by :

(a) 10-mesh screen (b) **12 mesh screen**

(c) 14-mesh screen (d) Sieving

76. Instantization means :

(a) Instant drying of milk (b) **Instant soluble**

(c) Readily consumable (d) For solution making

77. NIRO drying system is a :

(a) British system (b) **Danish systems**

(c) American system (d) German System.

78. Agglomeration means :

(a) Getting together

(b) **Adherence of powder particle**

(c) Seperation of particles

(d) Detaching of particles

79. Instantization is basically applied for :

(a) WMP (b) Malted Food

(c) **SMP** (d) Baby Food

80. Wetting is done by :

(a) Water (b) Steam

(c) **Steam & atomized water** (d) Milk

81. Cyclone is associated with :

(a) SCM (b) USCM

(c) Condensing (d) **Instantization**

82. Storage temperature for milk powder is :

(a) 10°C (b) 15°C

(c) 5°C (d) **Less than 24°C**

83. Yield of milk powder depends mainly on :
(a) Moisture (b) Protein (c) **TS** (d) Fat

84. Size of powder particle under drum drying system is :
(a) 10-250 mm (b) **8-20 mm**
(c) 6-12 mm (d) 10-12 mm

85. Bulk density of spray dried powder is :
(a) **0.5-0.6** (b) 0.4-0.5 (c) 0.3-0.5 (d) 0.2-0.3

86. Colour of spray dried cow milk WMP is :
(a) Yellow (b) **Light yellow**
(c) Whitish (d) Darker

87. Flavour of drum dried powder is :
(a) Like milk (b) Like baby food
(c) **Definitely cooked** (d) Like instant milk powder

88. Calandria is assocaited with :
(a) Baby food (b) WMP
(c) SMP (d) **SCM**

89. Cyclone unit is basically used for :
(a) Condensing (b) **Separation in drying**
(c) Blowing (d) Collecting

90. Powder is cool at :
(a) **room temperature** (b) 20°C
(c) 15°C (d) 10°C

91. Stale flavour in powder is due to :
(a) Decomposition (b) **Long storage**
(c) Poor milk (d) Improper packaging

92. Ransidity in powder is due to :
(a) High heating (b) Fat decomposition
(c) **Low preheating** (d) Sunlight

93. Tallowy defect is due to :
    (a) Decomposition (b) Lipase
    (c) **Sunlight** (d) Moisture
94. Lumpiness in powder is due to :
    (a) **Insufficient drying** (b) overheating
    (c) Excess fat (d) Excess protein
95. Caked defect in powder is due to :
    (a) Low heating (b) high heats
    (c) **Moisture absorption** (d) Bad packagings
96. SMP is mainly used for :
    (a) Food manufacture (b) **Recombined milk**
    (c) Curd (d) Ice-cream.
97. WMP is mainly used for :
    (a) **Reconstitution** (b) Sweets
    (c) Sweet curd (d) Bread
98. WMP has a shorter shelf-life, because of :
    (a) Protein (b) Lactose
    (c) **Fat** (d) Moisture
99. Browning in powder is due to :
    (a) Short storage (b) **Long storage**
    (c) Lactose (d) Protein
100. Atomizer is used for :
    (a) **Feeding the dryer** (b) Cooling
    (c) Seperation (d) Moisture

Glossary

Age thickening : The gradual increase in viscosity, due to physico-chemical changes, which occur during storage of condensed milk. The thickening is, sometimes, caused by micro-organisms producing acid or rennet like enzymes.

Agglomeration : Clustering, when milk powder particles collide amongst themselves and adhere to form clusters.

Atomization : It is the process of the bulk break-up of liquid into millions of individual droplets, forming a spray.

Bacteriocidal : Capable of killing bacteria.

Bacteriostatic : Capable of inhibiting or slowing down the growth and reproduction of bacteria.

Bulk Density : It is the weight of a unit volume of powder; in practice, it is expressed as g/ml, g/100 ml or g/l when powder is shipped over long distances, it is important that it has a high bulk density as the reduced volume sales on transportation cost and packaging low bulk density is also an important characteristic of instant powder.

Casein : Constitutes about 80% of the total protein present in milk. It is phospho-protein responsible for the viscosity and white colour of milk. In buffalo milk, casein content is about 3.5% and in that of cow milk 3.0%.

Chhana : Chhana is an indigenous milk product obtained by acid coagulated of hot milk followed by draining of whey. Chhana is a product of great commercial importance base for Bengali sweets such as Rosogolla, Sandesh, Rajbhog, Chhana podo etc.

CIP : Cleaning-in-place means that rinsing water and detergent solutions are circulated through tanks, pipes and process line without the equipment having to be dismantled. CIP can be defined as circulation of cleaning liquids through machines and other equipments in a cleaning circuit.

Cyclone : Cyclones are centrifugal particle separators. In the cyclone, air with powder particles at a high velocity is introduced tangentially into a cylindrical vessel which has a much larger cross section than the entering duct. The velocity is decreased in the cylinder, thus, permitting the settling of the solids. The velocity of the air decreases near the wall of the cyclone and the product falls by gravity and is removed from the bottom.

Dispersability : It is referring to powder's ability to disperse (spread out) in an appropriate amount of water to prevent humping.

Drying : It means that the water in a liquid product (in case of milk, concentrate) is removed, so that the product takes on a solid form. Drying extends the shelf-life of the milk, simultaneously reducing its weight and volume. This reduces the cost of transporting and storing the product.

Emulsifier : A substance that make possible uniform distribution of fat in dairy products. It improves the whipping ability of the mix and its texture. Only emulsifiers specified under rules 60 to 62 of the PFA can be used in dairy products.

Evaporation : In the dairy industry evaporation means the concentration of liquid milk, products containing dissolved, emulsified or suspended constituents. During this process water is removed by boiling.

Forewarming : The milk is preheated to 80°C/no hold for destruction of pathogens, other harmful microorganisms and to inactive enzyme. The choice of temperature employed for forewarming tends to vary depending upto the microbial quality of milk as well as reconstitution and sensory characteristics desired in the final product.

Freeze concentration : It is a relatively new method for removing water from aqueous solution. Basic principle is cooling the solution to temperature below its freezing point to produce ice crystals and by removing ice crystal, TS of content is increased.

Freeze-drying : In this process, the product is deep-frozen and the water frozen in the product is evaporated under vacuum. It is also called sublimation drying. The most important advantages of freeze drying is the greatest possible degree of preservation of the structure, the colour, the vitamin and aminoacid content and the flavour and odour of the freeze dried product.

Head space : The space in a container between the surface of a food and the underside of the lid.

Heat Classification of Skim milk powder : The powder is classified according to the heat treatment to which the raw milk has been subjected before converting it to a powder. The heat classification is expressed indirectly by the level of undenatured whey protein in the powder.

Heat Stability : The resistance of casein particles to aggregation (coagulation) during heat processing is usually referred to as heat stability.

Heat Sterilisation : Destruction of the majority of micro-organisms in a food by heating.

Hermetically-sealed Container : A package that is designed to be secure against entry of microorganisms and maintain the commercial sterility of its contents after processing.

Homogenize : To break-up fat globules of milk by forcing through the minute opening under pressure in order to make the particles uniformly small and evenly distributed.

Immunoglobulin : One of a group of proteins (globulins) in the body that acts as antibodies.

Infant milk foods : Standardized milk dried after addition of specified carbohydrates (cane sugar, dextrose and dextrine, maltose or lactose) iron salts and vitamins. It should not contain starch or anti-oxidants. Moisture content 3.5-5.0%, milk fat 18-28.0%, total carbohydrates not less than 20.0% minerals upto 8.5%.

Instantization : The process by which dried milk and milk products are made instantly soluble by solubilization or rehydration.

Instantizing : A process which imparts good rewetting properties to a dry product is called instantizing.

Instant-milk powder : Both skim milk and whole milk powder with extremely good wettability and solubility, known as instant milk powder. Instant powder dissolves instantly even in cold water.

Khoa : Khoa refers to the important traditional milk product of India prepared by partial dehydration of milk by heating under controlled conditions. Khoa is a product of great commercial importance base for the preparation of a variety of indigenous milk sweets such as Burfi, Peda, Milkcake, Gulabjamun etc. throughout the country.

Lactoferrin : It is an iron binding glycoprotein that occurs in cow milk at a level of 0.2 mg/ml. Mother's milk contains large amounts of lactoferrin (0.1-0.2 g/100ml). It plays an important role in the resistance against intestinal infection particularly *Escherichia coli*.

Lactose : The principal sugar of the milk of all mammals. Cow milk contains 4.5-5.0 per cent lactose. Relatively less soluble than cane sugar. On fermentation gives rise to lactic acid. Improper crystallization of lactose leads to 'Sandiness' in sweetened condensed milk.

Lamination : Bonding together of two or more packaging films, papers or foods.

Lecithins : A phospholipid mixture of a diglycerides of fatty acids; used for emulsifying, dispersing agents and as antioxidant.

Maillard reaction : Maillard browing occurs when proteins interact with reducing sugar. In Maillard reaction, lactose and lysine residues in milk proteins (mainly casein) are reactants. The early Maillard reaction is confined to the formation of the protein-bound Amadori product lactuloslysine. In the final stage, brown pigments (melanoidins) are formed and protein polymerisation occurs. This is the major problem of UHT sterilized milk during storage.

Malted milk : A product obtained by combining milk with the extract from a mash of ground barley-malt and wheat flour, with or without the addition of salt sodium and potassium bicarbonate, cocoa powder, starch may be absent. Contains 3.3% moisture, 14.6% protein, 21.5% lactose, 8.5% milk fat, 1.5% maltose and 47.8% dextrine.

Mold button : Mold buttons are lumps of variable size, cheese consistency and whitish yellow to reddish brown colour. These buttons are found on the surface or in the subsurface layers of the sweetened condensed milk.

Multi-effect evaporator : It is an assembly of evaporators which designed in such a way that the vapours from one effect are guided to the 2nd effect to be used as steam and so on, thereby increasing the steam economy.

Nutritive value : It is the nutritional status of a feed or the percent availability of nutrients from a feed.

Operation Flood : World's biggest dairy development programme aimed to modernise India's dairy Industry was launched in July 1970 to 'Flood' the four metropolitan cities from eighteen 'Anand pattern' dairy cooperatives in their hinterland milkshed areas. The Rs. 116 crore integrated project which ended in June 1981 was designed and executed by the National Dairy Development Board (NDDB) and the Indian Dairy Corporation (IDC).

Oxidation : A process involving addition of oxygen, loss of hydrogen; also a reaction involving the loss of electron.

Oxidized flavour : Defect in milk due to fat oxidation.

Rancidity : A flavour defect in milk due to fat oxidation and hydrolysis of fat by lipases.

Relative humidity : The ratio of the partial pressure of water in air to the pressure of saturated water vapour at the same temperature, mutiplied by 100.

Reverse Osmosis (RO) : Reverse osmosis is a dewatering technique. RO membranes separate solutes with molecular weight of approximately 150 daltons and above. Hence, fat, proteins, lactose and all undissociated minerals are retained and concentrated by the membrane and only water and some ionized minerals are allowed to pass through.

Roller drying : It is based on drying preconcentrated milk on the surface of a hot drum. As the water evaporates, a thin layer of dried milk is formed on the drum that is continuously removed by a scraper. The product is in the form of scales, which are later pulverized in a hammer mill.

Score card : A numerical standard of excellence with definite values assigned for different aspects; a basis of judgement of a product.

Seeding in condensed milk : Seeding refers to the introduction of lactose in a very fine powdered form during the cooling process to provide nuclei for crystallisation. The purpose of seeding is to give the lactose present in the sugar saturation form, an additional incentive for crystallization.

Sinkability : It is referring to powder's ability to sink in appropriate amount of water after becoming wet.

Skim milk powder : Made by removing most of the fat and water from milk. The moisture content must not exceed 5% and the fat content should not exceed 1.5%.

Solids-not-fat (SNF) : The percentage of total solids in dairy products minus the fat percentage. Solid-not fat present in milk means the total solids minus butter fat.

Solubility : It is referring to the properties of the powder in connection with the formation of a solution or a stable emulsion.

Solubility Index of milk powder : It is the volume of sediment in millilitres, which is found at the bottom of the centrifuge tube. It is a reverse scale, i.e., greater the solubility of the powder, the lower is the solubility index. High milk acidity, prolong heating, high drying temperature, high moisture content and ageing of sample effect the solubility index of skim milk powder.

Soyabean milk : An imitation milk prepared from soyabean.

Spray drying : In spray drying, drying depends upon converting the liquid into an exceedingly fine mist in which the droplets are about 0.002 inch in diameter, which evaporates spontaneously by contact with a current of hot air. The temperature of hot air may be between 100-150°C. The fine dust of dried milk can be removed mechanically.

Tallowiness : A flavour defect in butter due to oxidation of unsaturated fatty acid, catalysed by traces of copper and oxidizing enzymes. It may be caused by exposure to air, light or heat, metals and metallic salts.

Total Cell Count (Standard plate count) : The total number of bacteria or fungi both viable and non viable in a sample of milk or dairy product; the number of organisms growing under specific laboratory conditions from one millilitre of milk.

Water activity : The ratio of vapour pressure of water in a solid to that of pure water at the same temperature.

Weaning food : It is the special protein rich food for the weaned child.

Wettability : It is referring to powder's ability to become wet when transferred to a water surface of appropriate size.

Whey Protein : Whey is made up of different protein fractions. There are five major fractions and some minor fractions. The main fractions of whey protein are β lactoglobulin, α-lact albumin, Bouine serum albumin, Immunoglobulins, Glyco macropeptide. The minor fractions of whey protein are Lactoferrin, Lactoperoxidase and lysosime. Nutritional value of whey protein is very high which consists of biological value 104 and PER value 3.6.

Whey-Protein Nitrogen Index : Heat treatment denatures whey proteins, the percentage denatured increasing with the intensity of the heat treatment. The degree of denaturation is normally expressed by the whey protein Nitrogen Index (WPNI) i.e., in milligrams of undernaturated. Whey protein nitrogen per gram of powder. (mg/g powder)

Whole milk powder : When water is removed from milk, dried whole milk is the resultant product; such a product must have at least 26% milk fat and not more than 5% moisture.

Collateral Readings

1. Abbot, J., Glover, F.A., Nuir, D.D. and Skudder, P.J. (1979). Application of reverse osmosis to the manufacture of dried whole milk and skim milk, *J. Dairy Res.*, 46, 663.

2. Abd-El-Salam, A.M., Hafi, A.A., Mahran, G.A and Abdel Hamid, L.B., (1973). *J. Sci Food Agri.*, 24, 1163.

3. Agricultural and Processed Food Products Export Development Authority (APEDA).

4. American Dry Milk Institute Inc. Chicago (1971), Standards for Grades of Dry Milk including method analysis, Bulletin 916.

5. Anantakrishnan, C.P., Khan, A.Q. and Padmanabhan (1993), *Milk products preparations and Quality Control,* Shri Lakshmi Publication, Madras, 131-184.

6. Andersen, M. (1989), Improvements of powder characteristics, In. Food Ingredients Europe, Conference proceedings Maarssen, Netherlands', Expoconsult, Publishers, 139-140, Damrow Co., Denmark.

7. Aneja, R.P., Mathur, B.N., Chandan, R.C. and Banerjee, A.K. (2002), Technology of Indian Milk products, A Dairy India Publication, Delhi India. 177, 180-181.

8. Anon (1983), The Prevention of Food Adulteration Act, 1954 and PFA Rules, 1955, pp. 215-216.

9. Anonymous (1988), "*Dairy Handbook*" Alfa Laval, Lund Sweden, pp. 301.

10. Anonymous (1991), Natural Cheese Powder, *Food Process* 52(11) 114, 116.

11. Anonymous (1993), A savoury assortment, *Dairy Ingredients Int.* 58(8) 30-31.

12. Arora, K.L (1987), Quality of Sweetened Condensed Milk Manufacture in India, *Indian Dairyman,* 39 (9) 415-417.

13. Backe, A.G., Newstead, D.T. and Sanderson, W.B. (1974), XIX Int. Dairy Congress, IE, 614.

14. Balasubrahmanyam, N. (1995), Plastic in Food packaging in Profile on Food Technology, Ministry of Food processing Industries, Central Food Tech. Res. Inst. Mysore, and Indian Inst. of Packaging, Bombay, p. 24.

15. Banerjee Animesh, (1999), Export potential of Indian Dairy products. *Indian Dairy Man,* 51 (4) 19-22.

16. Barbano, D.M., Bynum, D.G. and Senyk, G.F. (1983), Influence of reverse osmosis on milk lipolysis, *J. Dairy Sci.*, 66, 2447.

17. Beck, B. (1971), Siisswaren, (1920) 964, 966, 968.

18. Beever, P. and Growhurst, D., Fire and explosion hazards accociated with spray drying operation (1989), *J. of the Soc. of Dairy Technol.* 42 (3) 65-70.

19. Beever, P. (1984), Spontaneous ignition of milk powders in a spray-drying plant, *J. of the Soc. of Dairy Technol.* 37 (2) 68-71.

20. Beever, P. (1985), Fire and explosion hazards in the spray drying of milk, *J. of Food Technol.*, 20, 637-645.

21. Beever, Paula (1984), *Journal of Society of Dairy Technology,* 37, 2, 68-71.

22. Bell, E.F. (1989), Upper Limit of Vitamin E, in infant formulas, *Journal of Nutrition*, 119, (12S) 1829-1831.

23. Bennett, R.J. (1988), Alternatives to evaporation. In evaporation and spray drying in the Newzealand Dairy Industry. Paper presented at a workshop Palmerston North new Zealand, 9th May, 1988, Palmerstion North New Zealand, Food Technology Department Massey University

(1989) 33-44.

24. Bhadania, A.G., Patel S.J. and Shah, U.S. (1986), Use of roller dryer for manufacure of Khoa, *Indian J. Fd. Sci. Technol*, 23(4) : 295-299.

25. Bhanumurthi, J.L. (1977), Formulated products need for New product development and process innovations. A lecture delivered at the summer institution on Technical problems in the manufactue of dairy products under Indian conditions, NDRI, Karnal.

26. Bhattacharya, D.C., Mathur, O.N., Tiwari, B.D., Balachandran, R. and Srinivasan, M.R. (1971), Studies on the method of production of Shrikhand and Shrikhand powder, Paper presented at the 8th Dairy Industry Conference, Trivandrum.

27. Bindra, H. (1979), Nutritional evaluation of cow and buffalo milk based infant food formulae, M.Sc. thesis, Kurukshetra University, Kurukshetra, India.

28. Biston, H. (1983), *Int. Dairy Fed. Bull. Doc.*, 157, p. 3.

29. Blanc, B. (1981), Biochemical aspects of human milk and Comparison with bovine milk, *World Rev. Nutr. & Dietet.*, 36 : 1-89.

30. Blanchfield, J.R. (2000), Food labelling, Wood head Publishing Cambridge.

31. Blissett, D.R. (1969), *Dairy Inds.*, 34(10) 633.

32. British Standards Institution (1971), P.d. 6459, Guidance on avoiding odour from packaging materials used for food stuff London.

33. Brydson, J.A. (1969), Plastic Materials, London, Iliffe Books Ltd.

34. Buchheim, W. and Dejmek, P. (1990), "Milk and Dairy Type Emulsions", In Food Emulsions 2nd. Ed. K. Larsson and S.F. Friberg. eds. Marcel Dekker, New York, pp. 203-246.

35. Buhler M and Olofsson, M. (1983), US patent, 4388329.

36. Buma, T.J. (1971), *Neth. Milk Dairy,* J. 25, p. 3.

37. Buma, T.J. (1971), *Neth. Milk Dairy,* J. 25. 107.

38. Bweton, H. (1988), Ultra-High Temperature Processing of Milk and milk products, Elsevier London, 354 pp.

39. Campey, D.R. (1987), The application of laser marking In : A Turned

(ed) Food Technology International Europe Sterling Publications. Interhnational, London, pp. 303-304.

40. Caric, M. (1994), Concentrated and dried dairy product jointly published by VCH Publishers, Ltd., U.K., VCH vertagsges Uchaft mbH, Germany and H. Publishers, Inc. N.Y.

41. Causeret, J. (1977), *Ann. Nutr. Aliment,* 25, A. 313-A, 334.

42. Chander Mohan (1977), Utilization of Soybean and whey for the manufacturing of weaning food of high protein efficiency ratio, *Ph.D. Thesis*, Punjab University, Chandigarh.

43. Chendrasekhara, M.R., Shurpalekar, S.R., Ray B.H.S., Kurien, S. and Shurpalekar, K.S. (1966), *J. Food. Sci. Technol.* 3, 94.

44. Cheryan, M. (1985), Membrance processing. In the *Year Book of Science and Technology,* Mc-Graw Hill Book Company, N.Y., p. 182.

45. Cheryan, M. (1998), Ultrafiltration and Microfiltration Hand book. Technomic Publishing Co., USA.

46. Condex Alimentarius Commission (1981) International Standards for Foods for Infants and children, FAO, Rome, Italy.

47. Coulter, S.T. and Jenness, R. (1964), Dry Milk Products, Chapter 20 in Food Dehydration ed. Van Arsdel W.B. and Copley M.J. AVI Publishing Co., Inc. West Port, C.T.

48. Darrington, H. (1999), The power of powder. *Food Manuf.,* 74 (March) 39-40.

49. David, J. (2005), Technological Advances in Indigenous Milk Products, Kitab Mahal, Allahabad.

50. De, A.K. and Patel, R.S. (1989), Technological aspects of manufacturing Chakka powder, *Jap. J. Dairy Fd. Sci.,* 38 : 169-173.

51. De, S. (1976), Dried Chhana, *Indian Dairyman,* 28 (9) 389.

52. De, Sukumar (1980), Outlines of Dairy technology Oxford University Press, Reprint 2001, 268-381.

53. Deboer, R. and Nooy, P.F.C. (1980). Concentration of raw whole milk by reverse osmosis and its influence on fat globules. Destination, 35, 201.

54. Dhillon, L.S. (2005), Manufacturing of malt-milk based food products. Indian Dairman, 57(4) 59-66.

55. Dixon, B.D. (1985), Dairy Products prepared from reverse osmosis concentrate market milk products, butter, skim milk powder and Yoghurt *Aust. J. Dairy Technol.*, 40, 91.

56. Donnelly, J.K., O'Sullivan, A.C. and Delaney, RAM (1974), Reverse Osmosis Concentration Application, *J. Soc. Dairy Technol.*, 27, 128.

57. Drew, P.G. and Manners, J.G. (1985), Microbiological aspect of reverse osmosis concentration of milk, *Aust. J. Dairy Technol.*, 40, 108.

58. Encyclopaedia of Packaging Technology (1986), T.E. Wolmsley and M Baker (editors). John Wiley & Sons. N.Y.

59. Fellows, P.J. (2000), Food processing technology—principles and practice 2nd. Woodhead Publishing, United Abington Hall, Abington. Cambridge CBI, 6aH, England.

60. Filka, P. (1984), Safety aspects of spray drying. Cited in Drying 84 Ed. Arun S. Majumdar, pp. 296-301.

61. Food and Drug Administration (1991) Code of Federal Regulations, *Food and Drugs,* Part 107 – Infant Formula 107, 100 Washington D.C. 79 pp.

62. Ganguli, N.C. and Kuchroo, C.N. (1979), *Indian Dairyman,* 31. 691.

63. Ghosh, B.C., Rajorihia, G.S. and Pal, D. (1986), Complete gulabjamun mix. *Indian Dairyman,* 38 (6) : 279-282.

64. Ghosh, B.C. and Kulkarni's (1990), Dehydrated Khadi : A Convenience food. *Indian Dairyman,* 42(1) : 11-14.

65. Ghosh, B.C., Rajorihia G.S. and Pal, D. (1984), Formulation and storage studies of gulabjamun mix powder, *Indian J. Dairy Sci.,* 37 (7) : 362-367.

66. Glicksman, M. (1975), Carbohydrates for fabricated foods, In fabricated foods, AVI Pub. Co. Westport, USA, pp. 68-88.

67. Glover, F.A. (1971), Concentration of milk by ultra filtration and reverse osmosis, *J. Dairy Res.*, 38. 373.

68. Glover, F.A. (1985), Ultrafiltration and Reverse Osmosis for the Dairy Industry, *Technical Bulletin 5, National Institute for Research in Dairying, Reading,* U.K. 207 pp.

69. Glover, F.A. (1986), Modifications to the composition of milk in Modern Dairy Technology, Vol. 1 R.K. Robinson, ed. Elsevier, London, pp.235-272.

70. Golani Rajesh (2005), Developments in Milk Powder Technology, *Indian Dairyman* 57(12) 70-73.

71. Goldman, A.S. and Smith, C.W. (1973), *J. Pediatr.*, 82, 1082.

72. Goyal, G.K. and Paltani, I.P. (2003), Consumer packaging of Dairy products and Safety, *Packaging India*, 35 (6) 19-27.

73. Guinee, P.P., O'Brien, N.B. and Rawle, D.F. (1983), Development of a range of cheese powder products for USC in cheese sauces Proc. 2nd Dairy Levy Symposia, Keogh, K.K. ed. Teagase. *Dairy Products Research Centre*, Moore Park, Fermoy Co. Cork, Ireland, pp. 17-42.

74. Guinee, T.P. and Kikawley (2004), Cheese as an ingredient, in Cheese : Chemistry, Physics and Microbiology, 3rd ed. vol. 2. Major cheese group. Edited by Patrick F. Fox, Paul L.H., McSweeney, Timothy Mcogan and Timothy Plwinec. Elsevier Academic Press, 84, Theobald's Road, London. WC1X8RR, U.K., pp. 415-418.

75. Guinee, T.P., O'Brien, N.B. and Rawle D.F. (1994), The viscosity of cheese sauces with different starch systems and cheese powders, *J. Soc. Dairy Technol.,* 47, 132-138.

76. Guise, B. (1987a), Spotlight on Sachet packaging, *Food Process,* 56, 35-37.

77. Guise, B. (1987b), Filling an industry need, *Food Process July*, 31-33.

78. Gupta, Vijay Kumar. (2005), Role of Membrance processing in value additiion of dairy product in Souvenir, National Seminar on value added dairy product, at NDRI, Karnal, Dec., 21-22, pp. 118-121.

79. Gupta, S.K. and Arora, A. (1982), Development of Jowar-Soybean – Skim Milk Based Wearing food, *Indian J. Dairy Sci.*, 35 (1) 62-67.

80. Gupta, S.K. and Patel, A.A. (1978), *Indian Dairyman,* 30, 713-715.

81. Gupta, S.K., Patil, G.R. and Patel, A.A. (1987), Fabricated Dairy Products. *Indian Dairyman,* 39 (5), 199-208.

82. Gurr, M.I. (1981), *J. Dairy Res.*, 48, 519.

83. Gyorgy, P. (1953), *P. Diatrics,* 11, 98.

84. Hall, C.W. and T.I. Hedrick (1975), Drying of Milk and Milk products, The AVI Pub. Co. Inc. Connecticut.

85. Hallstrom, B. (1985), 'Heat Exchange' in Evaporation, Membrane Filtration and Spray drying in Milk Powder and cheese production, R. Hansen, ed. *North European Dairy Journal*, Vanlose Denmark, pp. 25-54.

86. Hansen, S.O. and Hansen, P.S. (1990), Spray dried whole milk powder for manufacture of milk chocolate, Scandinavian Dairy Information, 4 (2) 79-82.

87. Harding, F. and Gregson, R (1978), Nitrates and Nitries in skim milk and whey powder in the United Kingdom, *Proceeding of the XXth International Dairy Congress*, pp. 356-357.

88. Harper, W.J. and Hall, C.W. (1976), Dairy Technology and Engineering A.V.I. Pub. Co. West Port Connecticut.

89. Hassan, H.N., Mistry, V.V. (1990), High protein lactose free milk powder Ii. Functional, Chemical and physical properties (abstracts), *Journal of Dairy Science,* 73 (Suppl. 1) 114.

90. Hedrick, T.L. (1981), Spray drying of cheese, *Pro. Second Biennial Marschall International Cheese Conf.*, Madison w. 1, pp. 16-80.

91. Heiss, R. & Eichener, K. (1971), Fd. Mf. 46.5.

92. Hersom, A.C. and Hulland E.D. (1980), Canned Foods, 7th edn. Churchill Livingston, London.

93. Hiddink, J., Boer Rde and Nooy P.F.C. (1980), Reverse Osmosis of Dairy Liquids, *J. Dairy Sci.,* 63, 204.

94. Hulbert, T. Lacy (1970), Modern packaging of milk products, Journal of the Society of D. Tech., Vol. 23, No. 2, 100-103.

95. Hunziker, O.P. (1949), Condensed Milk and Milk Powder, The Author La Grange Illinois.

96. Indian Standards (1968), IS 1547, Specification for infant milk foods. Indian Standard Institution New Delhi, India.

97. Indian Standards (1975), IS. 1165, Specifications for Milk powder, Indian Standards. Institute, Nanak Bhawan, New Delhi.

98. Indian Standards (1977), IS 1224 (Part I), Determination of fat in whole milk. Indian Standards Institute, Nanak Bhawan, New Delhi.

99. Indian Standards (1985) IS : 1547, Specification for infant milk food (2nd revision), Indian Standards Institute, New Delhi.

100. Indian Standards (1986), Specification for condensed milk, partly skimmed and skimmed condensed milk, 2nd Revision, IS : 1166.

101. International Dairy Federation Bulletin (1987), Recommendations for fire prevention in spray drying of milk powder, No. 219, pp. 1-15.

102. International Dairy Federation, IDF Standard 87 (1979) Square Vergote 41, 1040 Bruxelles.

103. Jha, Alok (2000), Development of process for longlife kheer and Instant kheer mix, Ph.D. Thesis. National Dairy Research Institute, Karnal.

104. Jha, Alok and Mann, R.S. (1996), Fire and Explosion hazards in the spray drying of milk, *Indian Dairyman,* XLV 111 (5) 23-26.

105. Jha, Alok and Patel, A.A. (2002), Instant Kheer mix—A value added convenience product for Indian dairy industry, *Indian Dairyman*, 54 (10) 55-59.

106. Joglekar, S.S. (1984). Selective aspects of manufacturing and improved infant formula, Ph.D. thesis, Kurukshetra University, Kurukshetra, India.

107. Johnson and Alford (1987), Fundamentals of dairy chemistry, CBS Publishes & Distributors.

108. Joshi, A.A. (1987), Packaging of dairy products. In Dairy India. R.R. Gupta Publishers, New Delhi, 44-48, 69.

109. Kansal (1992), Lactose in human health. *Indian Dairyman,* 44 : 497-500.

110. Keele, C.A. and Neil E. (1966), Sampson and Wright's Applied physiology, The English language book society, Oxford University Press, U.K.

111. Kessler, H.G. (1981), Food Engineering and Dairy Technology, Published by Werlag A. Kessler Pub., Germany.

112. Kessler, H.G. (1988), Food and Biotechnology, *Dairy Technology*, 3rd ed. A Kessler, Freising, Germany, 582 pp. (in German).

113. Khanna, A.K. (2005), Fortification of milk & Dairy products with vitamin A, *Indian Dairyman,* 57 (7) 31-37.

114. Kinby, C.J., Brooker, B.E. and Law, B.A. (1987), *Int. J. Food Sci. Technol.,* 22, 355-375.

115. King, L. (1999), New cheeses for the new millenium. *Food Manuf.* 74 (Sept.), 39-40.

116. Knipschildt, M. (1986), Drying of Milk and Milk Products In Modern Dairy Technology, Vol. 1 R.K. Robinson ed. Elsevier London, pp. 131-234.

117. Knipschildt, M.E. and Anderson, G.G. (1994), Drying of milk and milk products in Modern dairy technology, Vol. I. Edited by R.K. Robinson. Chapman & Hall, 1994.

118. Kosikowski, F.V. and Mistry, V.V. (1997), Drying and Freezing of cheese in cheese and Fermented Milk food vol. Origins and Principles, Kosikowski, F.V. and Mistry V.V., eds. F.V. Kosikowski L.L.C. West port CT., pp. 318-327.

119. Kuchroo, C.N. and Ganguli, N.C. (1980), Technological approaches for infant food manufacture from modified buffalo milk, *J. Food Sci.,* 45 : 133-1335.

120. Lampert, L.M. (1970), Modern dairy products, Chem. Pub. Co., New York.

121. Langherdries, J.P., Detry, J., Van Hees, J., Lamboray, J.M. Darimont, J. Mozin M.J., Secretin M.C. and Senterre J. 1995, Effect of a fermented infant formula containing viable bifidobacteria on the faecal flora composition and pH of healthy full-term infants, J. *Paediatr Gastroenterol Nutr.* 21 : 125-129.

122. Lea, C.H. & Hannan R.S. (1952), *Fd. Sci.,* 228.

123. Lea, C.H.; Maran T & Smith, J.A.B. (1943), *J. Dairy Res.,* 13, 162.

124. Lewin, L. (1996), Powder puff, *Dairy Indus. Int.,* 61(7) 35.

125. Lovell, H.R. (1972), Packaging of dried milk product, *Journal of the Society of Dairy Tech.,* Vol. 25. No. 3, 148-157.

126. Lyster, R.L.J. (1970), *J. Dairy Res.,* 37, 233.

127. Mahajan, B.M. (1971), Studies on the production and Shelf-life of spray dried Shrikhand, M.Sc. thesis, Panjab University Chandigarh.

128. Marshall, S.C. (1985), The reverse Osmosis Plant Operation, *Aust. J. Dairy Technol.*, 40(3) : 88-90.

129. Marshall, S.C. (1985), The reverse osmosis process, *Aust. J. Dairy Technol.*, 40, 86.

130. Master, K. (1976), Spray Drying, George Godwin Limited London.

131. Master, K. (1979), Spray Drying Hand Book, John Wiley and Sons, New York.

132. Masters, K. (1985), Preventing fire/explosion occurrence or damage in open-cycle spray dryers. Cited in *Spray Drying Handbook*, pp. 365-372.

133. Mayer, J.B. (1969), *Phys. Med. Rehabilitation,* 10, 16.

134. Mendez Seijido, C. (1989), Identification & management of hazards in powdered products including babyfood : HACCP indicator organisms of keeping quality. In Modern Microbiological methods for dairy products. Bru sets, Belgium; *International Dairy Federation,* Special Issue No. 8901, 106-112.

135. Mettle, A.E. (1978), *J. Soc. Dairy Technol.*, 26, 84.

136. Mettle, A.E. (1980), *J. Soc. Dairy Technol.*, 33, 150.

137. Mil, P.J.J.M Van; Bouman, S. (1990b), Freeze concentration of dairy products, *European Dairy Magazine,* No. 1 31.

138. Mil, P.J.J.M. Van; Bouman, S. (1990), Freeze concentration of dairy products, *Netherlands Milk and Dairy Journal,* 44(1) 21-31.

139. Min. D.B., Ticknor, D.B. Lees, S.H., Reinecius, G.S. (1990), Effects of processing conditions and anti-oxidants on the oxidative stability and carbon-dioxide formation in low fat dry milk, *Journal of Food Science* 55(2) 401-403. 423.

140. Missel, D. (1996), Selecting the right ingredients for adding the flavour of cheese, *Food Product Design,* 6(5) 51-61.

141. Mistry, V.V.; Hassan, H.N. (1990), High protein, lactose free milk powders. I. Production and Composition. (Abstract), *Journal of Dairy Science,* 73 (Supplement 1) 113.

142. Mortensen, H. (1999), Cheese powder—a natural ingredient. *Scand Dairy Inform.*, (1) 14-15.

143. Mrithyunjaya, N. and Bhanumurthi, J.L. (1987), Shelf-life of laminate and tin packed whole milk powder, *Indian J. Dairy Sci.*, 40(1) pp. 78-81.

144. Munjial Manmohan (1995), Manufacture of heat classified powder, *Indian Dairyman,* 57(11). 19-25.

145. NDDB (1990), Milk Products manual, National Dairy Development Board, Anand, 46-86.

146. Niro Atomizer (1970), Danish Patent, 123067.

147. Niro Atomizer (1971), British Patent, 1301796.

148. Niro Atomizer (1973), U.S. Patent, 33773519.

149. Osborne, A (1986), Cracking Codes of communication. *Food Process,* January, 29-31.

150. Packaging Code, B.S. 1133.

151. Packard, V.S. (1982), Human Milk and Infant Formula, Academic Press. New York, 269 pp.

152. PAG (1971), Protein rich mixtures for use as weaning foods. Protein Advisory group of the United Nations Guidelines, No. 8, New York.

153. Pahwa, A. (1982), Studies on the role of bifidus factor in infant nutrition. Ph. D. Thesis, Kurukshetra University, Kurukshetra.

154. Pahwa, A. and Mathur, B.N. (1982), *Indian J. Dairy Sci.*, 36, 229.

155. Pahwa, A. and Mathur, B.N. (1982), Nutritional evaluation of a bifidus containing infant formula, *Int. Dairy Congr.,* IE ; 664.

156. Pahwa, A. and Mathur, B.N. (1987), Assessment of a Bifidus containing formula, Part I : Evaluation of the Biological Attributes, *Indian J. Dairy Sci.,* 40(2) 261-267.

157. Pahwa, A. and Mathur, B.N. (1987), Assessment of a Bifidus containing Infant Formula, Part II : Implantation of Bifidobacterium bifidus, *Indian J. Dairy Sci.,* 40(3) 364-367.

158. Paine, F.A. (1962), Fundamentals of Packaging, London : Blackie & Son Ltd.

159. Paine, F.A. (1971), *Fd. Trade Rev.,* 41, 46.

160. Pal D., Rajorhia, G.S. Garg, F.C. and Verma, B.B. (1994), Production of dried rasogolla mix from ultrafiltered milk retentate 24th Int., *Dairy Congr.*, Melbourne, Australia, p. 424.

161. Pal. D. and Cheryan, M. (1987a), Membrane Technology in Dairy Processing. Part I : Reverse Osmosis. *Indian Dairyman,* 39(6) : 251-267.

162. Pallansch, M.J. (1970), "Dried Product". In By-products from Milk B.H. Webb and F.O. Whittier eds. AVI Publishing Company, West port. CT., pp. 124-182.

163. Patil, G.R. and Singh, R.R.B. (2005), Current R & D in value Added Dairy Products in Souvenir. National Seminar on value added Dairy products at National Dairy Research Institute, Karnal, Dec. 21-22, pp. 36-43.

164. Paul SC and Mathur, B.N. (1993b), Effect of lactose hydrolysis on some nutritional characteristics of spray dried lactose hydrolysed infant formula, *Aust. J. Dairy Technol.*, 48 : 49-53.

165. Paul, S.C. (1984), Technological aspects of manufacturing low-lactose infant formula., Ph.D. Thesis, Kurukshetra University, Kurukshetra, India.

166. Paul, Sc. and Mathur, B.N. (1993a), Development of law lactose infant, *Infant J. Dairy Sci.,* 46 : 377.

167. Pedersen, P.J. (1985), "Plants for recombination", In Evaporation Membrane Filteration and Spray Drying in Milk Powder and Cheese Production. R. Hansen ed. North European D.J. Vanlose, Denmark, pp. 373-384.

168. Peebles D.D. (1958), U.S. patent, 2835586.

169. Peebles, D.D. (1936), U.S. Patent, 2054441.

170. Peebles, D.D. and Clary, D.D. Jr. (1955), US patent 2710808.

171. PFA 1999, Prevention of Food Adulteration Act (1954), and Rules (1955) as amended upto 1999. Law Publications, New Delhi.

172. Pisecky, J. (1990), *Scand Dairy Interm.,* 2/90 S. 336.

173. Pisecky, Jan. (1986), Standards, specifications and test methods for dry milk products. In : concentration & drying of foods. Edited by Diarmind Morecarthy, London, Elsevier Applied Science Publishers, 1986.

174. Planine, F. & Milohnoja, M. (1968), Proc. VI Int. Symp. on F.microbiol. 344.

175. Potter, N.N. (1968), Food Science., A.V.I. Pub. Co., Westport, Conn.

176. Prasad C, Prasad, D. (1990), Texture changes and their assessment in concentrated/evaporated milk during manufacturing and storage. *Indian Dairyman,* 42(6) 274-277.

177. Prasad, C. Prasad S; Singh, M.D. (1989), Physico-chemical changes in evaporated milk during manufacture & storage, *Indian Dairyman,* 41(10) 505-508.

178. Prasad, S. and Gupta, S.K. (1988), *Agric. Rev.,* 9, 81-104.

179. Raiha N.C.R. (1989), Protein Quality and Whey-Casein Ratio in Infant Formulas In Protein and non-protein Nitrogen in Human Milk S.A. Atkinson and B. Lonnerdal, eds. CRC press. Boca Raton FL, pp. 137-144.

180. Rajendran, S. Godavari, S. and Narsimhan, K.S. (1993), Fumigation of gulabjamun ready-mix, *J. Fd. Sci. Technol.,* 30(3) 226-228.

181. Rajhoria, G.S. (1994), Manufacture of Khoa powder in Advances in traditional dairy products, *CAS in Dairy Technology,* NDRI, Karnal, pp. 17-20, 107-110.

182. Ranganna, S. (2000), Packaging materials for processed foods in Hand book of canning and Aseptic Packaging. Tata McGraw-Hill Publishing Company limited, New Delhi, pp. 270.

183. Rao, B.V.R. (1985), Development of a formula for the dietary management of pre-term infants, Ph.D. thesis, Kurukshetra University, Kurukshetra, India.

184. Rao, B.V.R. and Mathur, B.N. (1987), Processing changes during Manufacture of a Spray dried Infant Formula : Part I. Electrophoretic and ion-exchange chromatographic behaviour of Casein, *Indian J. Dairy Sci.,* 40. 4, 398-403.

185. Rao, B.V.R. and Mathur, B.N. (1987), Development of a pre-term Infant Formula : Part I, Effect of forewarming Temperature on physical characteristics, *Indian J. Dairy Sci.,* 40 (2) 249-253.

186. Rao, B.V.R. and Mathur, B.N. (1987), Development of a Pre-term Infant Formula : Part II, Effect of foreworming Temperature on the Physical Changes during storage, *Indian J. Dairy Sci.,* 40(2) 254-260.

187. Rao, B.V.R. and Mathur, B.N. (1987), Processing Changes During Manufacture of Spray Dried Infant Formula : Part II. Gel Chromatographic and Electrophoretic Behaviour of Whey Proteins, *Indian J. Dairy Sci.*, 40(4) 404-409.

188. Rao, B.V.R. and Mathur, B.N. (1987), Processing Changes During Manufacture of Spray Dried Infant Formula : Part III. Interrelationship between Protein Interactions and Reconstitution characteristics, *Indian J. Dairy Sci.*, 40(4) 410-417.

189. Raychaudhuri, A.K. (1996), Dairy Development in India (A brief history), Part I, *Indian Dairyman,* 48(5) 17-22.

190. Retsina. T. (1988), Agglomeration : a process to improve fine powder handling. In *Food Technology International*, Europe, 37-39.

191. Rizvi, S.S.H., Mann, R.S.and Ali, A.L. (1987), A case study of appropriate technology transfer : development of automated, continuous khoa powder manufacturing process, *Indian Dairyman,* 31(1) 63-68.

192. Samel, R and Muers, M.M. 91962a). Age thickening in sweetened condensed milk, Part I, *J. Dairy Res.*, 29, 249-258.

193. Samel, R. and Muers, M.M. (1962c), Age thickening in Sweetened Condensed milk, Part III, *J. Dairy Res.,* 29, 269-277.

194. Samel, R. and Muers, M.M. 91962b). Age thickening in Sweetened condensed milk, Part II, *J. Dairy Rs*, 29, 259-267.

195. Sanderson, N.Z. (1967), *J. Dairy Technology*, 2, 35.

196. Sanyal, M.K. and Balachandran, R. (1989), Changes in certain properties of spray dried malted milk stored in polyethylene pouch and glass bottle, *Asian J. of Dairy Research*, 8(2) 75-80.

197. Sardana Vijay and Dokras Ashish (2005), Global Trends in Major Dairy Products Trade in Sourvenir. National Seminar on value added dairy products, *NDRI Karnal*, 21-22, Dec. pp. 15-25.

198. Sen, D.C. and Gupta, S.K. (1987), Effect of Freezing and Thawing on the Sensory Qualities of Dairy Products, *Indian Dairyman,* 39 (5). 231-234.

199. Singh, M.N. (1982), Studies on the modification of bovine milk for the development of an infant formula, Ph.D. thesis, Kurukshetra University, Kurukshetra, India.

200. Singh, M.N. and Mathur, B.N. (1982), Application of electrodialysis for mineral iodification of bovine mikl for the manufacture of infant formula, *Int. Dairy Congr.,* IE : 514.

201. Sirett, R.R., Eskrit, J.D. and Deriakta, E.J. (1981), U.S. Patent, 4264 638.

202. Skov, O. (1994), Protection against fire and explosion in spray dryers, Cited in Modern Dairy Technology, Vol. 1, Ed. R.K. Robinson, pp. 255-272.

203. Soneren T.H.M., Brinkhuis, J.A., Damman, A.J. and Klok, H.J. (1984), *Meths. Milk Dairy J.,* 38. 43-53.

204. Soneren, T.H.M., Damman, A.J. and Klok, H.J. (1982), The viscosity of skim milk concentrate, *Neths. Milk Dairy, J.* 36, 305-316.

205. Stabile R.L. (1983), Economics of reverse Osmosis and multistage evaporation for concentrating skim milk from 8.8 to 45% solids,. *J. Dairy Sci.,* 66. 1765.

206. Techline Ind. Data Services Ltd., U.K. (1998), Using less to do more. Tetra Pak Company magazine, No. 18. Tetra Pak Marketing Services, AB Lund, Sweden, p. 12.

207. Tetra Pak Processing System AB (1995), Dairy Processing Handbook, Tetra Pak Processing System, A.B. Lund, Sweden.

208. Thompkinson, D.K. (1984), Technological aspects of incorporating vegetable oil for infant food formulations, Ph.D. Thesis, Kurukshetra University, Kurukshetra, India.

209. Thompkinson, D.K.; Sabikhi Latha and Mathur, B.N. (2005), Infant Food Formulations : Status and Strategies in Indian Dairy Industry, Vol. 1, Dr. Chawla information centre, Pvt. Ltd.

210. Thompkinson, D.K. and De, S. (1981), Utilisation of dried Khoa for Peda making, *J. Fd. Sci. Technol.,* 18(4) 259-262.

211. Thompkinson, D.K. and Mathur, B.N. (1987), Modification of Buffalo milk lipid profile for Infant Food Formulation, *Indian J. Dairy Sci.,* 40(4). 418-421.

212. Tiwari, B.D. and De, S. (1976), Standardization of the Industrial method of production of Dried Chhana, *Indian J. Dairy Sci.,* 29 (3), 212.

213. Tran, K.M. and Einerson, M.A. (1989), U.S. Patent, 48, 57341.

214. Unnikrishnan, V. Bhavadasan, M.K. and Surendranath, B. (2003), A Drymix for convenient preparation of Palada Payasam, *Indian Dairyman* 55 (7) : 70-74.

215. Upadhyay, K.G. (1998), Age thickening in sweetened condensed milk, *Indian Dairyman*, 50 (10) 27-34.

216. Upadhyaya Ram, L. and Kilara Arun (1999), Two stage drying of Milk and Whey, *Indian Dairyman,* 51 (7) 7-10.

217. US Department of Agriculture (1976), Agriculture Research Service Agricultural Handbook, No. 8-1 (revised), Washington D.C.

218. Vanden Berg M.G. (1984), The Thermization of Milk IDF Bulletin Document, 182, 3.

219. Versteeg, C. (1985), Commercial reverse Osmosis plants, *Aust. J. Dairy Technol.,* 40, 118.

220. Warner. James N. (1976), Principles of Dairy processing Wiley Eastern limited, New Delhi.

221. Webb, B.H., Johnson A, Hand Alford, J.A. (1978), AVI Publication Company, INC, West port, Connecticut.

222. Westergaard Vage, Milk Powder Technology, NIRO A/S Denmark.

223. Westergard, V. (1983), Milk Powder Technology, Evaporation and Spray Drying, 3rd ed. NIRO Atomizer, Copenhagen Denmark, 147.

224. Wilcox George (1971), "Milk Cream and Butter Technology", Noyes Data Corp, USA.

225. www.usdec.org.

226. Zavarin, Y.A., Chekulaeva, L.V. and Chekulaev, N.M. (1978), *20th Int. Dairy Congress, Brief Communications,* Vol. E, p. 736.

Subject Index

P

R

S

T

V

W